T0332259

Video Shot Boundary Detection

RIVER PUBLISHERS SERIES IN INFORMATION SCIENCE AND TECHNOLOGY

Consulting Series Editor

KWANG-CHENG CHEN
National Taiwan University
Taiwan

Information science and technology enables 21st century into an Internet and multimedia era. Multimedia means the theory and application of filtering, coding, estimating, analyzing, detecting and recognizing, synthesizing, classifying, recording, and reproducing signals by digital and/or analog devices or techniques, while the scope of "signal" includes audio, video, speech, image, musical, multimedia, data/content, geophysical, sonar/radar, bio/medical, sensation, etc. Networking suggests transportation of such multimedia contents among nodes in communication and/or computer networks, to facilitate the ultimate Internet. Theory, technologies, protocols and standards, applications/ services, practice and implementation of wired/wireless networking are all within the scope of this series. We further extend the scope for 21st century life through the knowledge in robotics, machine learning, cognitive science, pattern recognition, quantum/biological/molecular computation and information processing, and applications to health and society advance.

- Communication/Computer Networking Technologies and Applications
- Queuing Theory, Optimization, Operation Research, Statistical Theory and Applications
- Multimedia/Speech/Video Processing, Theory and Applications of Signal Processing
- Computation and Information Processing, Machine Intelligence, Cognitive Science, and Decision

For a list of other books in this series, see www.riverpublishers.com

Video Shot Boundary Detection

Krishna K. Warhade

SPANN Lab, Department of Electrical Engineering,
Indian Institute of Technology Bombay, Mumbai 400076, India

Shabbir N. Merchant

Department of Electrical Engineering,
Indian Institute of Technology Bombay, Mumbai 400076, India

Uday B. Desai

Indian Institute of Technology Hyderabad,
Yeddumailaram 502205, Andhra Pradesh, India

Published 2011 by River Publishers
River Publishers
Alsbjergvej 10, 9260 Gistrup, Denmark
www.riverpublishers.com

Distributed exclusively by Routledge
4 Park Square, Milton Park, Abingdon, Oxon OX14 4RN
605 Third Avenue, New York, NY 10017, USA

Video Shot Boundary Detection / by Krishna K. Warhade, Shabbir N. Merchant, Uday B. Desai.

Routledge is an imprint of the Taylor & Francis Group, an informa business

ISBN 978-87-92329-71-4 (print)

Contents

Preface

The detection of shot boundaries provides a base for nearly all video abstraction and high-level video segmentation approaches. Therefore solving the problem of shot boundary detection is one of the major prerequisites for revealing higher level video content structure. Moreover, the other research areas which can be benefitted considerably from successful automation of shot boundary detection processes are distance learning, telemedicine, interactive television, digital libraries, multimedia news, video restoration and geographical information system. Despite all the research activity in shot boundary detection, there are some issues which have not been adequately addressed and need to be resolved. The major challenges for shot boundary detection are, detection of gradual transition and elimination of the disturbances caused by abrupt illumination change and motion. The disturbances caused by abrupt illumination change is mainly due to flashlight, fire, flicker and explosion, whereas the disturbance caused by motion is due to rapid camera and object motion in the consecutive frames. It is difficult to develop a single approach which is not only invariant to various disturbances mentioned above but should also be sensitive enough to capture the details of visual content and have excellent detection performance for all types of shot boundaries. In this monograph, we focus on detection of wipes and avoiding the disturbances due to flashlight, fire, flicker, explosion and motion.

Wipe transition detection in video segmentation is more difficult to detect than abrupt and other gradual transitions, due to diversity in patterns, and difficulty in distinguishing wipe from camera and object motion. An algorithm has been proposed for wipe transition detection. In the proposed algorithm, first the moving lines due to wipe are obtained, which helps in eliminating most of the edges due to object boundaries and retains true wipe boundaries. Then Hough transform is applied on these moving lines to detect and categorize various wipe types. In order to decrease the computational load of the proposed algorithm, preprocessing step has been proposed as a first stage of the algorithm. The preprocessing step consists of calculation of statistical image difference between the consecutive frames to obtain the potential wipe frames. These potential wipe frames are processed through the proposed algorithm to detect and identify wipes. The proposed algorithm detects and identifies various types of wipes and also distinguishes wipes from object and camera motion. Performance comparison of the proposed algorithm, with and without preprocessing, with the other existing techniques clearly exhibited its effectiveness in terms of better Recall, Precision, F1 measure and Detection Rate.

Elimination of disturbances caused by flashlights is one of the major challenge in shot boundary detection. The existence of flashlight changes the luminance and chrominance abruptly due to sudden visual effect across the video sequence. This contributes to a larger difference between consecutive frames and leads to false detection of shot boundary. Existing flashlight detection methods have limitation in terms of duration of flashlights, nature of flashlights, computation time and calculation of an appropriate global threshold. Global threshold is a fixed threshold and needs to be adjusted for each new video. To address these issues an algorithm has been proposed for shot boundary detection due to abrupt transition in the presence of flashlight. In the proposed algorithm, the illumination effect due to flashlight was suppressed using logarithmic transform followed by discrete cosine transform, and then discrete wavelet transform based metric was used to find potential shot boundaries, and finally, local or adaptive threshold was used to declare shot boundary. Local and an adaptive threshold depends on a mean and standard deviation of the feature difference metric within a temporal window and hence overcomes the limitation of a global threshold. The proposed algorithm is tested on movie videos, and experimental results validate the effectiveness of the method to avoid false positives due to flashlights in shot boundary detection.

Detection of fire in video for fire alarm systems has been studied by many researchers, but detection of shot boundaries under fire, flicker and explosion (FFE) is one of the under studied areas. In thriller movies, FFE occur more often than other special effects and lead to false detection of shot boundary. Major metrics have been tested for detection of shot boundaries under FFE for various movies. It has been observed that precision is low for almost all metrics due to false positives caused by FFE. We proposed an algorithm based on cross correlation coefficient, stationary wavelet transform, and combination of local and adaptive thresholds for detection of shot boundaries under FFE. The proposed algorithm is tested on three movies and experimental results validate the effectiveness of our method in terms of better Recall and Precision.

We evaluated the performance of traditional shot boundary detection metrics in the presence of camera and object motion in RGB, HSV and YUV color spaces for eight movies. It has been observed that all these metrics provided poor results due to disturbances caused by these motion. The maximum false positives and missed detections are due to frame difference between consecutive frames caused by fast camera motion.

Developing a single approach which will eliminate the disturbances due to illumination as well as motion is a challenging task. To address these challenge, we propose an algorithm for shot boundary detection in the presence of illumination and motion. In this proposed algorithm, extraction of structure feature of each frames were done using dual tree complex wavelet transform. These structure features were insensitive to illumination and motion between consecutive frames. Then spatial domain structure similarity algorithm was applied on these structure features to find potential shot boundaries, finally correct shot boundaries are declared by using local and adaptive thresholds. The performance comparison of the proposed algorithm with other existing techniques, validate its effectiveness in terms of better Recall, Precision and F1 score.

In this monograph, we provide an insight into the advances in shot boundary detection. The work presented in this monograph is the outcome of research conducted at SPANN Lab, Electrical Engineering Department, Indian Institute of Technology Bombay. The authors express their gratitude to Dr. Vikram Gadre for his kind cooperation and encouragement extended during this research. Krishna Warhade acknowledges the Chairman, Directors and Principal of Lokmanya Tilak College of Engineering for their support. Krishna is thankful to his parents Keshavrao and Sumitra Warhade, his wife Rupali, his daughters Bhargavi and Tanishka, other family members and colleagues for their continuous support, sacrifice, and motivation during the entire duration of research and completion of this book.

We are grateful to Dr. Rajeev Prasad, Mrs. Jolanda Karada and other members of the River Publishers team for their kind cooperation and support during the publication of this monograph. We also sincerely appreciate the efforts of Mr. Pratik R. Senghani for designing the cover picture.

List of Figures

List of Tables

Abbreviations

SBD	Shot Boundary Detection
1D	One Dimensional
2D	Two Dimensional
DCT	Discrete Cosine Transform
DWT	Discrete Wavelet Transform
DT-CWT	Dual Tree Complex Wavelet Transform
FFE	Fire Flicker Explosion
AT	Abrupt Transition
GT	Gradual Transition
R	Recall
P	Precision
F1	Harmonic mean of Recall and Precision
DR	Detection Rate
NFRWL	Number of frames with flashlights in the test video
SSIM	Spatial domain Structure similarity
CWT	Continuous Wavelet Transform
SWT	Stationary Wavelet Transform
CC	Cross correlation coefficient
CMS	Correlation of modified sign
LHR	Likelihood ratio
HD	Histogram difference
CS	Chi-square test
PD	Pixel differences
PPSSIM	Post processing on SSIM index
JPEG	Joint Photographic Experts Group
MPEG	Moving Picture Experts Group
TRECVid	Text REtrieval Conference series on Video retrieval evaluation
HVS	Human visual system
RGB	Red Green Blue color space
HSV	Hue Saturation and Value color space
YIQ	Luma in-phase quadrature color space

YUV	Luminance Y and two color difference U, V color space
SVM	Support Vector Machine
HMM	Hidden Markov Model
SWIII	Star War III movie
SWI	Star War I movie
JA	Jodhaa Akbar movie
BN	Bhooth Nath movie
SH	Sleepy Hollow movie
ID	Independence Day movie
DR	Deep Rising movie
SWIII	Star War III movie
PH	Pearl Harbor movie
TM	The Marine movie
SPR	Saving Private Ryan movie
XM	X-Men movie
HA	Home Alone movie
MI	Mission Impossible movie
JMP	Jumper movie
WED	Wednesday movie
PR	Pale Rider movie
BEE	Bee movie
GBU	Good Bad and Ugly movie

1

Introduction

1.1 Motivation

Shot boundary detection is the most basic temporal video segmentation task. It provides a basis for nearly all video abstraction and high-level video segmentation approaches. Therefore solving the problem of shot boundary detection is one of the major prerequisites for revealing higher level video content structures. Moreover, other research areas also can benefit considerably from successful automation of shot boundary detection processes. However, the detection of gradual transition, the elimination of disturbances caused by abrupt illumination change and large object/camera motion, have found to be the major challenges to the shot boundary detection. Hence it is necessary to develop an algorithm which is not only invariant to various disturbances mentioned above but also has an excellent detection performance for all types of shot boundaries. We took these challenges to the shot boundary detection as an opportunity to propose algorithms which will address these specific issues.

In our research, we primarily focus on the detection of wipes (gradual transition) and eliminating the disturbances caused by to flashlight, fire flicker and explosion, object motion, and camera motion.

1.2 Introduction to Video Shot Boundary Detection

Recent advances in multimedia compression technology, coupled with the significant increase in computer performance and the growth of the internet, have led to the widespread use and availability of digital video. The rapidly expanding applications of video have spurred the growing demand of new technologies and tools for efficient indexing, browsing and retrieval of video data. The increasing availability of digital video has not been accompanied by an increase in its accessibility. This is due to the nature of video data, which is unsuitable for traditional forms of data access, indexing, search and retrieval, which are either text based or based on the query-by-example paradigm. Therefore techniques have been sought that organize video data into more compact representations or extract semantically meaningful information. Such an operation can serve as a first step for a number of different data access tasks, such as browsing, retrieval, genre classification and event detection. Shot boundary detection is the most basic temporal video segmentation task, as it is intrinsically and inextricably linked to the way that video is produced. It is a natural choice for segmenting a video into more manageable parts, and thus it is very often the first step in algorithms that accomplish other video analysis tasks.

The development of shot boundary detection algorithm has the longest and richest history in the area of content based video analysis and retrieval. This area was actually initiated about a decade ago by the attempts to detect hard cuts in a video, and a vast majority of all works published in this area so far addresses in one way or another the problem of shot boundary detection. The detection of shot boundaries provides a basis for nearly all video abstraction and high-level video segmentation approaches. Therefore solving the problem of shot boundary detection is one of the major prerequisites for revealing higher level video content structure. Moreover, other research areas have profited considerably from successful automation of shot boundary detection processes such as distance learning, telemedicine, interactive television, digital libraries, multimedia news, video restoration, and geographical information systems.

The video streams are formed by editing different video segmentation known as shots. A shot is a sequence of frames generated during a continuous operation and it represents continuous camera action in time and space. Shots can be joined together in either an abrupt transition mode or gradual transition mode. In the abrupt transition two shots are simply concatenated, whereas in gradual transitions, additional frames may be introduced using editing operations such as dissolve, fade-in fade-out and wipe. The changes between shots can belong to the following categories:

- Abrupt transition or hard cut: This is the classic abrupt change case, where one frame belongs to the disappearing shot and the next one to the appearing shots as shown in Figure 1.1.
- Dissolve transition: In this case, the last few frames of the disappearing shot temporally overlap with the first few frames of the appearing shots. During the overlap, the intensity of the disappearing shot decreases from

Figure 1.1 Abrupt transition

normal to zero (fade-out), while that of the appearing shot increases from zero to normal (fade-in) as shown in Figure 1.2(a).

- Fade transition: Here, first the disappearing shot fades out into a blank frame, and then the blank frame fades in into the appearing shot as shown in Figure 1.2(b).
- Wipe transition: This is actually a set of shot change techniques, where the appearing and disappearing shots coexist in different spatial regions of the intermediate video frames, and the region occupied by the former grows until it entirely replaces the latter as shown in Figure 1.3(a).
- Digital video effects transition: There is a multitude of inventive special effects used in motion pictures. These are in general very rare and difficult to detect as shown in Figure 1.3(b).

1.3 Major Challenges to the Shot Boundary Detection

To achieve satisfactory detection performance, special attention has to be paid to deal with several challenges to the shot boundary detection. The detection of gradual transition and the elimination of disturbances caused by abrupt

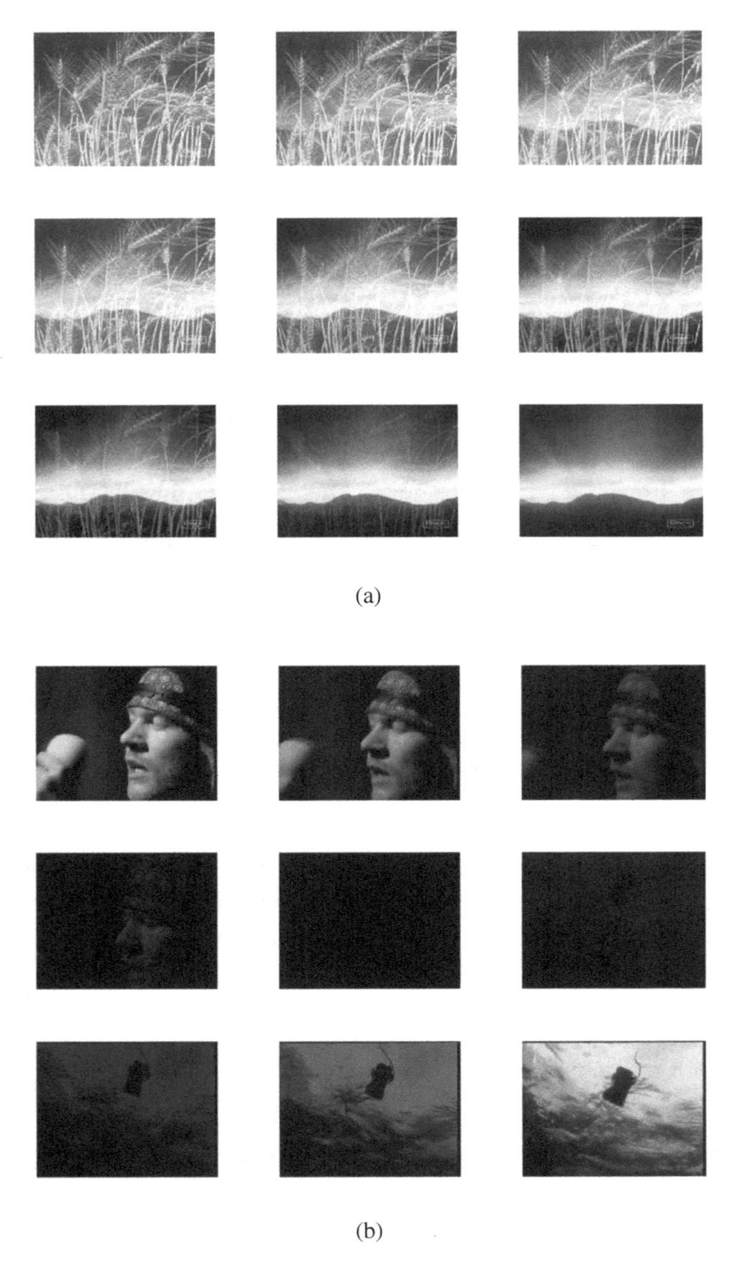

Figure 1.2 (a) Dissolve transition. (b) Fade transition

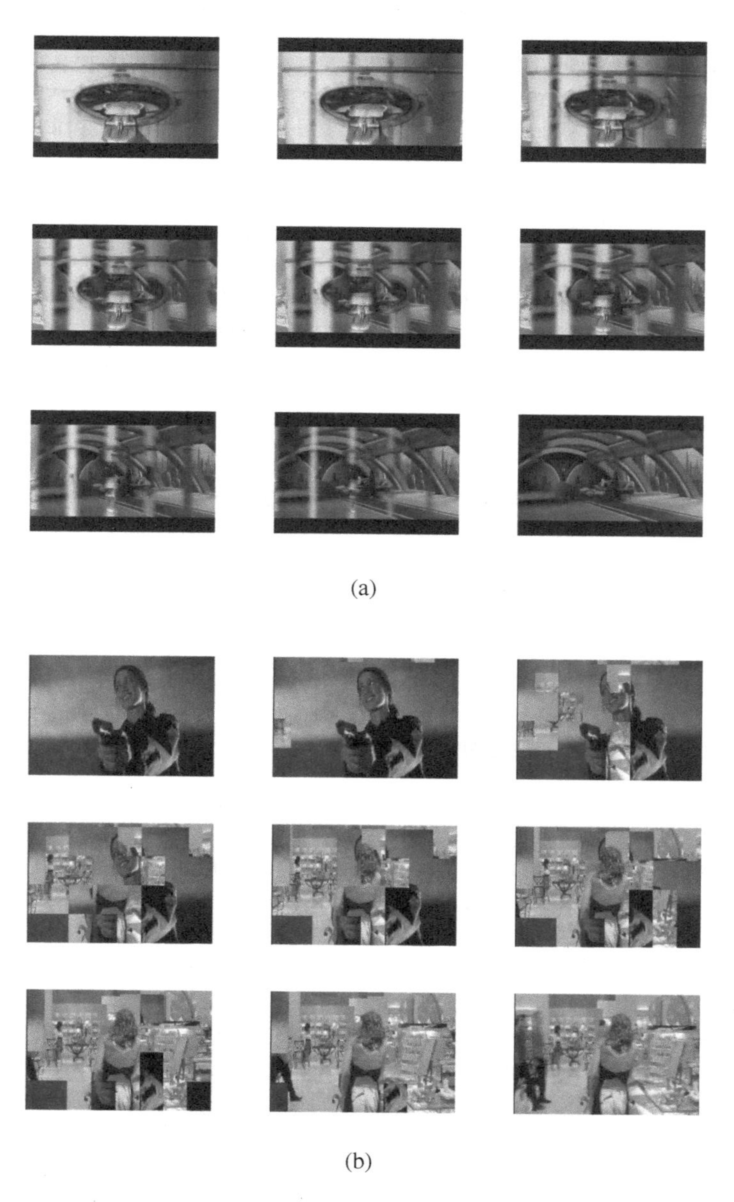

Figure 1.3 (a) Wipe transition. (b) Digital video effects transition

illumination change or large object/camera movement, have been found as the major challenges to the shot boundary detection [3, 5–12].

- *Detection of gradual transitions*: Detection of gradual transition is more difficult than that of abrupt transition as it include various special editing effects such as dissolve, wipe and fade resulting in a distinct temporal pattern over the continuity signal curve. The temporal patterns due to gradual transition are similar to those caused by object/camera movement, since both of them are essentially processes of gradual visual content variation.
- *Elimination of disturbances due to abrupt illumination change*: Most of the content representation methods are based on the color feature, in which luminance is a basic element. Abrupt illumination changes, such as flashlights, fire flicker and explosion often cause significant discontinuities of inter-frame feature, which is often mistaken for shot boundaries.
- *Elimination of disturbances due to large object/camera movement*: Besides shot transitions, object/camera movements also lead to the variations of visual content. Sometimes, the abrupt motion will cause similar continuity values to those of hard cuts, and the persistent slow motion will result in temporal patterns over continuity signal curve similar to those of gradual transition.

1.4 Problems Addressed

In this monograph, we focus on the solution to the major challenges in shot boundary detection. Despite all the research activity in shot boundary detection, there are some issues that have not been adequately addressed, and which need to be resolved. The major challenges for shot boundary detection are detection of gradual transition and elimination of the disturbances caused by abrupt illumination change and motion. The disturbances caused by abrupt illumination change is mainly due to flashlight, fire, flicker and explosion, whereas disturbance caused by motion is due to rapid camera and object motion in the consecutive frames. It is difficult to develop a single approach which is not only invariant to the various disturbances mentioned above but also sensitive enough to capture the details of visual content and having excellent detection performance for all types of shot boundaries. We consider the major challenges to the SBD as the problems in our research and the solutions to these problems are discussed below.

- Wipe transition has received less attention from the research community [13–18] compared to dissolve, fade-in and fade-out transitions [19–23]. Wipe transition has complexity and diversity in its transition pattern and it is difficult to find out unique model for the wipe transition. Hence it is a challenge to identify the presence of all different types of wipe transition and identify their individual patterns. The existing methods suggested for wipe detection have the following major limitations. They are limited to the detection of a primary pattern, like vertical, horizontal, and diagonal wipes with only one or two moving lines and are unable to detect nonlinear and nonrigid changing patterns and boundary. Also camera and object motion and lines formed by moving object cause missed detection and false alarm. The existing methods also failed if the wipe border is blurred and when exiting and entering scenes both have dark background. Hence we propose an algorithm for wipe transition detection. In the proposed algorithm, first the moving strip due to wipe is obtained, which eliminates most of the edges due to object boundaries and retains true wipe boundaries, followed by the application of the Hough transform to these moving lines to detect and categorize the various wipe types.
- Another major challenge to the shot boundary detection algorithm is to eliminate the disturbances caused by to flashlight, which is one of the major sources of false positives . The existence of flashlight changes the luminance and chrominance abruptly due to sudden visual effects across the video sequence. This contributes to the larger difference between consecutive frames and leads to false detection of shot boundary. Histogram based metrics and edge detection based metrics are mostly used for shot boundary detection in the presence of flashlight [2, 24–26]. These are sensitive to strong lighting changes but are invariant to small object motion. Edge detection based methods are computationally expensive and they fail when objects are not clear due to dark background. Thus existing flashlight detection methods have a limitation in terms of duration and nature of flashlight, computation time, and threshold. To address these issues, an algorithm has been proposed for shot boundary detection in the presence of flashlight. In the proposed algorithm, the illumination effect due to flashlight is suppressed using logarithmic transform followed by discrete cosine transform, and then a discrete wavelet transform based metric is used to find potential shot boundaries and, finally, a local or adaptive threshold is used to declare shot boundary.

- Detection of fire in videos for fire alarm systems has been studied by many researchers [27–31], but detection of shot boundaries under fire, flicker and explosion (FFE) is one of the under-studied areas. In thriller movies, FFE effects occur more often than other special effects and lead to a false detection of shot boundary. Hence the detection of shot boundaries in thriller movies under FFE effects is a difficult task. The major problem arises from the fact that changes in luminance are not uniform over consecutive frames, as the direction and position of fire moves in the consecutive frames. In a few cases, appearance and disappearance of fire in the consecutive frames contribute to large differences. We have proposed an effective algorithm based on cross correlation coefficient, stationary wavelet transform and a combination of local and adaptive thresholds for shot boundary detection.
- Another major problem with shot boundary detection algorithms is avoiding the influence of fast camera and object motion. Most of the review papers have mentioned that elimination of fast camera and object motion is one of the main concerns for effective shot boundary detection [6,10–12,32]. Algorithms suggested for suppressing the influence of motion or making them insensitive to object and camera motion have the limitation that they are effective only if motion is slow [33–37]. These algorithms failed when camera and object motions are fast. To address this challenge, we propose an algorithm for shot boundary detection in the presence of motion. We develop the structure feature using dual tree complex wavelet transform and demonstrated its invariance to the disturbances caused by illumination and motion. In the proposed algorithm, initially the structure feature of each frame is extracted using dual tree complex wavelet transform, then potential shot boundaries are detected by applying a spatial domain structure similarity algorithm to the structure feature of consecutive frames, and finally correct shot boundaries are declared by using local and adaptive thresholds.

1.5 Contributions of the Monograph

The contributions of this research are as follows:

- We propose an algorithm for wipe transition detection. In the proposed algorithm, first the moving strip due to wipe is obtained, and then a Hough transform is applied to these moving lines to detect and categorize various wipe types. Applying the proposed algorithm to the whole video to detect wipes is computationally expensive. In order to decrease the computational load of the proposed algorithm, we propose a preprocessing step as the first stage of the algorithm. The potential wipe frames are obtained using the preprocessing step, and then processed through the proposed algorithm to detect wipes. We extensively tested our proposed algorithm, with and without preprocessing, on several movies like Star Wars III, Star Wars I, Jodhaa Akbar, Bhooth Nath and Tomcat. Experimental results are thoroughly evaluated using the performance metrics Recall, Precision, F1 measure, and Detection Rate and compared with the existing methods discussed in [13, 14, 16]. We extensively varied the values of the tuning parameter in the preprocessing step and tested the results on various movies for Recall and Precision. The effect on the performance due to the variation of these parameters has been investigated and the results are reported for Jodhaa Akbar, Tomcat, BhoothNath, and Star War III movies. Our proposed algorithm achieved a relatively better trade off between Recall and Precision compared to other algorithms. We tested the proposed approach on 31 different wipe effects including 18 complex and special wipes. The proposed method successfully avoids false positives caused due to object and camera motion, by differentiating wipe transitions and motion using various gradient patterns.

- We have presented an effective method for shot boundary detection in the presence of flashlights. The proposed algorithm suppresses the effect due to flashlight by using logarithmic transform followed by discrete cosine transform. Discrete wavelet transform based metric in combination with local and automatic threshold is used to find shot boundaries. We extensively tested our proposed algorithm on the movies Sleepy Hollow, Independence Day and Deep Rising, and experimental results are evaluated using the performance metrics Recall, Precision and F1 measure. The proposed method is also tested for shot boundary detection under various condition of flashlight such as weak flashlight, when flashlights are present in the shot boundary, when flashlights are present for long duration with switch-on/off of a flashlight and when objects are not clear during flashlights. We have also shown that the first stage of our proposed method can be a preprocessing step for other shot boundary detection methods [4] and can reduce the false positives due to flashlight in this method. Comparison results with other shot boundary detection methods proposed in [1–4] in the presence of flashlight for various movie sequences show the effectiveness of the proposed algorithm. The proposed method outperforms other compared methods in terms of trade-off between Recall and Precision.

- Detection of shot boundaries in the thriller movies under FFE effects is a difficult task. We have evaluated the effectiveness of various metrics discussed in [2, 3, 5, 38–40] and cross correlation, under FFE effects. It has been found that color ratio histogram and cross correlation coefficient metric perform better than the other compared metrics under these effects. Also the precision result is low in every metric due to the influence of fire flicker and explosion. The behavior of FFE under which these metrics failed has been discussed. We have proposed an effective algorithm based on cross correlation coefficient, stationary wavelet transform and a combination of local and adaptive thresholds for shot boundary detection. Experimental results on the movies Pearl Harbor, The Marine and Saving Private Ryan show significant improvements in terms of better Recall and Precision compared to existing methods proposed in [2, 3, 5, 38–40].
- Disturbances caused by fast object and camera motion are often mistaken as shot boundaries and its elimination is the major challenge to the shot boundary detection algorithms. We address this issues by developing structure features which are invariant to illumination and motion using dual tree complex wavelet transform. Then a spatial domain structure similarity algorithm which is also invariant to average luminance and contrast is applied to these structure features to detect shot boundaries. Finally local and adaptive thresholds are used to declare correct shot boundaries. We extensively tested our proposed algorithm on the action movies X-Men and Home Alone, where a large number of frames with fast object and camera motion is observed in addition to frames with illumination. Experimental results are thoroughly evaluated using the performance metric Recall, Precision, and F1 measure. Our proposed algorithm achieved a relatively better trade-off between Recall and Precision with high F1 measure compared to other tested algorithms proposed in [2, 35] on the same video data. The proposed algorithm is successful in avoiding disturbances due to illumination change and fast motion when the camera follows the object.

1.6 Organization of the Monograph

The monograph is structured as follows.

In Chapter 2, survey papers on the comparison of major shot boundary detection algorithms are discussed. Also a detailed survey of papers on wipe detection and shot boundary detection in the presence of flashlight, fire flicker explosion, and motion is been provided. The feature used for video frame representation and the major metrics used in shot boundary detection have been presented. Finally various decision methods and evaluation criteria for shot boundary detection have also been discussed in detail.

In Chapter 3, we discuss the proposed algorithm for wipe transition detection. A preprocessing step, as the first stage of the algorithm has been proposed to decrease the computational load of the proposed algorithm. The potential wipe frames obtained after the preprocessing step are processed through the proposed algorithm to detect wipes. The variation of the tuning parameter in the preprocessing step and its effect on results in terms of Recall and Precision are also discussed in this chapter. The experimental results of our proposed algorithm, with and without preprocessing, on various movies and comparison with various existing wipe detection methods are discussed in detail as well.

In Chapter 4, we present an effective method for shot boundary detection in the presence of flashlights. Experimental results with other shot boundary detection methods in the presence of flashlight for various movies are also discussed.

In Chapter 5, we evaluate and discuss the effectiveness of various metrics in the presence of illumination due to fire flicker and explosion effects. The behavior of FFE under which these metrics failed is also discussed. We propose an effective algorithm for shot boundary detection in the presence of fire flicker and explosion. Experimental results on movie data are being compared with the major shot boundary detection methods and discussed at a later stage in this chapter.

In Chapter 6, the performance evaluation of traditional shot boundary metrics in the presence of motion in various color space is presented. We also discuss the proposed algorithm for shot boundary detection in the presence of illumination and motion. It has been extensively tested on action movie data and the results are being compared with existing techniques.

We summarize our results and point to some directions for future work in Chapter 7.

1.7 Summary and Conclusions

In this chapter, we discussed the motivation behind this work and the basics of shot boundary detection. The major challenges to the shot boundary detection and contribution of our research are also discussed.

2

Overview of Approaches for Video Shot Boundary Detection

2.1 Literature Survey of Existing Methods in Shot Boundary Detection

Developing techniques for detecting shot boundaries in a video have been the subject of substantial research over the last decade. In this chapter we give an overview of the relevant literature. Initially, literature on comparative investigations on early shot boundary detection algorithms is discussed. Then we discuss the existing methods in the specific areas such as shot boundary detection in the presence of flashlight, fire flicker and explosion, camera motion and object motion, and finally existing algorithms to detect dissolve fade-in and fade-out are also discussed.

2.1.1 Comparison of Major Shot Boundary Detection Algorithms

A comparison of various shot boundary detection metrics has been reported by Nagasaka and Tanka [5]. They experimented with various metrics such as difference of gray-level sums, sum of gray-level differences, difference of gray-level histograms, colored template matching, difference of color histograms and χ^2 comparison of color histograms. They concluded that the most robust method is the χ^2 comparison of color histogram [5]. Their method is robust against zooming and panning of the camera, but failed to detect special effects such as fading. A simple and effective method for shot boundary detection is proposed by Zhang et al. [3]. A twin comparison approach has been developed to solve the problem of gradual transition. A motion analysis algorithm has been applied to eliminate the false interpretation of camera movement as transitions. Pair-wise comparison, likelihood ratio and histogram comparison are used as a different metric for video partitioning [3]. Zhang et al. observed that object motion, camera motion, flashing light and flickering objects are the major source of false positives. Hsu and Harashima [41] formulated the scene changes and activities as a motion discontinuity. Shahraray [42] proposed an algorithm to detect abrupt and gradual scene changes based on motion-controlled temporal filtering of the disparity between consecutive frames. Sethi and Patel [39] proposed the use of Kolmogorov–Smirnov test based on maximum absolute value difference between cumulative distribution functions. Sethi and Patel used only the DC coefficients of the I frame to perform hypothesis testing using luminance histogram. It is assumed that a separation between two I frames is fixed and small. The exact location of abrupt changes cannot be located with this method. Hamapur et al. [43] developed model based methods to capture the different types of shot transitions. According to Zabih et al. [44] hard cuts, fades, dissolves and wipes exhibit a characteristic pattern in the edge change ratio time series. The consecutive frames are compared for the number and position of edges that enter and exit between the two frames. Shot boundaries are detected by looking for large edge change percentages. Dissolves and fades are identified by looking at the relative values of the entering and exiting edge percentage. Boon-Lock Yeo and Bede Liu [45] proposed several rapid scene analysis algorithms for detecting scene changes and flashlight scenes directly on the compressed video. These algorithms operate on the DC coefficients which can be readily extracted from compressed video using Motion JPEG or MPEG without full-frame decomposition.

Boreczky et al. [6] presented a comparison of several shot boundary detection and classification techniques. Histograms, edge tracking, discrete cosine transform, motion vector and block matching methods are used for comparison. They concluded that algorithm features that produced good results are region based comparisons, running differences, and motion vector analysis. A combination of these features produced better results than region histogram or the running histogram algorithms alone. Ahanger et al. [7] discussed existing research trends in the area of segmentation and indexing of digital video. An overview of video indexing followed by a discussion on video segmentation using pixel based, likelihood ratio, histogram comparison, twin comparison, DCT coefficient and detection of object and camera motion are presented. Idris et al. [46] reviewed the image and video indexing techniques in detail. Various methods of video segmentation in the uncompressed and compressed domain are also discussed. The discussed methods in uncompressed domain are intensity/color based techniques, Histogram based techniques, block based techniques and twin comparison, whereas the methods discussed in a compressed domain include DCT-coefficient based methods, motion vector based methods, and hybrid DCT-Motion vector based methods. Lienhart [8,32] tested various existing shot detection algorithms on a diverse set of vide sequences. The evaluation is focused on the detection and recognition of hard cuts, fades and dissolve. Four shot boundary detection algorithms based on color histogram differences, edge change ratio, standard deviation of pixel intensities and contrast change are compared. They concluded that the performance of shot boundary algorithm based on edge change ratio is computationally expensive and its performance is inferior to that of methods based on color histogram differences, standard deviation of pixel intensities and edge based contrast. They showed that all the detection algorithms are influenced negatively by global and local motion in the video.

Gargi et al. [9] evaluated and characterized the performance of a number of shot detection methods using color histograms (in various color spaces), MPEG compression parameter information and image block-motion matching. They concluded that color-histogram based shot detection performs well at a moderate computational cost. Histogram intersection in the *Munsell* space had the best performance. Block-motion matching algorithms do not perform as well as color histograms or MPEG based methods, in addition to being computationally intensive. Ford et al. [47] provided a comprehensive quantitative comparison of metrics that have been applied to shot boundary detection in digital video sequence. They reported results on various histogram test statistics, statistic based metrics, pixel differences, MPEG metrics and an edge based metrics. They concluded that the Kolmogorov–Smirnov test is the best histogram metric; it performs better than the more popular chi-square and histogram difference metrics. Histogram metrics produce the best results when computed for blocks, rather than globally. The overall best results for abrupt cut detection are obtained by using the statistic based metrics computed at the block level. Hanjalic [10] identified and analyzed the major issues related to shot boundary detection problem in detail. Then conceptual solution to the shot boundary detection problem is provided in the form of statistical detector that is based on minimization of the average detection error probability. The required statistical function is modeled using a robust metric for visual content discontinuities, which includes the knowledge of the short length distribution, visual discontinuity pattern at shot boundaries and characteristics temporal change of visual features around a boundary. They found that the influence of motion and lighting changes on the detection performance cannot be reduced easily.

Becós et al. [48] critically reviewed most of the approaches used for shot boundary detection. They proposed a unified detection model centered on mapping the space of inter-frame distances onto a new space of decision better suited to achieving a sequence-independent thresholding. This mapping aims to consider frame ordering information within the thresholding process; it is based on the parametric modeling of the patterns that transitions generate on the distances output. Cotsaces et al. [49] reviewed a certain basic information extraction operation that can be performed on video. Their review is focused on shot boundary detection, video abstraction, and video summarization. Yuan et al. [11] conducted a comprehensive review of the existing approaches and identified the major challenges to the shot boundary detection. They proposed a general framework for shot boundary detection and presented a unified shot boundary system based on graph partition model and support vector machine classifier. They found that the detection of gradual transition, the elimination of disturbances caused by abrupt illumination change or large object/camera movement are the major challenges to the current shot boundary detection techniques. Smeaton et al. [12] presented an overview of the TRECVid shot boundary detection task, a high-level overview of the most significant of the approaches taken and a comparison of performances for seven years of TRECVid activities . Smeaton also gave a brief review of the state of art of video analysis indexing and retrieval [50], and provided promising research direction. He also discussed the challenges that researchers working in the area face.

2.1.2 Detection of Wipes

Alattar [13] developed a model for wipe detection, which exploits the linear change in the means and variance of frames in the wipe region. Wu et al. [51] proposed a wipe detection method based on the standard deviation of projected pixel-wise differences from DC images. Fernando et al. [14] applied the Hough transform to the image boundary to analyze wipe pattern. Kim et al. [52] introduced the concept of visual rhythm , which is an abstraction of the video. They have shown that different video effects manifest themselves as different patterns on the visual rhythm.

Drew et al. [15] detected wipe effects by investigating the orientation of boundary lines in spatial temporal images. Pei et al. [53] proposed the use motion vectors to find the scene change regions of each frame. Campisi et al. [54] developed an algorithm based on the trajectory estimation of the boundary line between two successive frames for wipe detection.

Han et al. [55] used 3D wavelet transform, Gaussian weighted Hausdorff distance, and the direction of motion vector to identify wipes. Nam et al. [16]

used wavelet transform to find direction-emphasized images. These images are projected on a specified angle by a Radon transform and then a B-Spline interpolation is used to identify wipe region. Mackowiak et al. [56] employed motion activity, and dominant color descriptors to find wipe region. Iwamoto et al. [17] used a HSV histogram to find potential wipe candidates, and extracted boundary lines from these candidates using the Canny edge detector and the Hough transform. Finally optical flow vectors of the multiple line are calculated to differentiate between object boundary and image boundary. Li et al. [18] proposed properties of independence and completeness to characterize ideal wipe. The Bayes rule is applied to each potential wipe to estimate an adaptive threshold for the wipe verification.

Li Yufeng et al. [57] proposed a method where color sub-image and edge sub-image of each frame are decomposed by wavelet. Then potential wipe frames are detected by the dynamic threshold and the Hough transform.

2.1.3 Shot Boundary Detection in the Presence of Flashlights

Methods for detecting or differentiating flashlight from abrupt transition can be broadly classified as histogram based methods, edge detection based methods and local window based methods.

Li and Lu [4] proposed a method to differentiate between shot and flashlights based on difference in area of background in edge image. Weixin et al. [24] used color ratio histogram as illumination invariant metric and local adaptive thresholds have been used to find shot change. Histogram difference is used as a metric to differentiate between abrupt transition and gradual transition [1], and then average intensity difference is used to detect flashlights. Model for flashlight detection is formed by length, strength, brightness, velocity, and impact of the flashlight [58]. Depending on these parameters thresholds are used to detect flashlight. In [25], Guimaraes et al. showed that flashlight effect forms long white vertical stripes in the visual rhythm by histogram image and these lines are extracted through a white top hat by reconstruction process.

In [26] edge direction, edge position, and edge matching are used to discriminate abrupt scene change from flashlight scenes. Yuliang and De [59] used twin comparison algorithm to find potential shot change. By using edge direction histogram and inter-frame similarities, light change and shot boundary can be detected. Accumulated histogram difference and energy variation is used to identify flashlights in [60]. In [2] frame difference from consecutive frames is detected using subdivided local color histogram comparison.

This difference is compressed by using logarithmic transform for efficient detection of flashlights.

2.1.4 Shot Boundary Detection in the Presence of Fire Flicker and Explosion

Alboil et al. [40] proposed a technique to detect abrupt transitions, when random brightness variations are present in the scene. This algorithm for shot boundary detection in the presence of flicker is a pixel difference based method, and is sensitive to the disturbances caused by fire flicker and explosion. Marbach et al. [27] proposed an algorithm which is based on the temporal variation of fire intensity captured by a visual image sensor. A candidate flame region is obtained by analyzing full image sequences. Then the fire features are extracted from this region to determine the presence of fire patterns.

Toreyin et al. [28] proposed an algorithm to detect flame and fire flicker by analyzing the video in the wavelet domain. This algorithm have used color and temporal variation information and checks flicker in flames using 1D temporal wavelet transform and color variation in fire-colored moving regions using 2D spatial wavelet transform. Celik et al. [29] developed a real-time fire-detector, which combines color information with registered background scene. They generated an adaptive background model of the scene using three Gaussian distributions for each channel. Then an adaptive background subtraction algorithm is used to extract foreground information and verified by the statistical fire color model to determine whether the detected foreground object is a fire candidate or not. This process forms the fire detection system and has been applied for the detection of fire in consecutive frames of video sequences.

The algorithm proposed by Paulo et al. [30] analyzed the frame-to-frame change in given features of potential fire regions. These features are color, area size, texture, etc. The change of each of these features is evaluated and the results are combined according to Bayes classifier to decide presence of fire. Celik and Demirel [31] proposed an algorithm where YCbCr color space have been used to separate the luminance from the chrominance more effectively than the other color spaces. Then rule based generic color model is used for flame pixel classifications.

The method proposed in [40] can be used to detect shot boundaries in the presence of fire flicker and explosion, whereas algorithms in [27–29, 31] can be used to detect fire in the fire detection systems and cannot be used for detection of shot boundaries. These algorithms can be used for detection of fire in movies and video databases, as well as for real-time detection of fire in surveillance systems. The method proposed in [30] is suitable for retrieval of fire catastrophes in newscast content, but cannot be used for detection of shot boundaries.

2.1.5 Shot Boundary Detection in the Presence of Motion

Lawrence et al. [33] proposed an algorithm for shot boundary detection based on the video's first-order partial derivatives. They tried to make algorithm insensitive to object and camera motion by considering areas with low apparent motion. Su et al. [34] proposed an efficient method for detecting dissolve type gradual transitions. They modeled a dissolve based on its nature and filtered out possible confusion caused by the effect of motion via statistics. They used binomial distribution model to distinguish real dissolves from motion.

Xu et al. [35] proposed a motion suppression technique for shot boundary detection based on 3D wavelet transform. Motion is characterized in terms of energy and variance, and then motion suppression value is extracted from intensity of motion energy. This motion suppression value is used to suppress motion influence suffered by traditional shot boundary detection algorithm. Jang et al. [36] proposed a shot transition detection method that compensates motion in a video. They first extracted motion vectors from two consecutive frames by using the size-variable block matching and then adaptive robust estimation method is used to estimate the global motion and eliminate local motion.

Park et al. [37] proposed two-motion based features for shot boundary detection of fast motion sequences. The first feature which is robust to fast motions is developed by considering the motion magnitude. The next feature, blockwise motion similarity, is defined by comparing motion directions in consecutive frames.

2.1.6 Detection of Dissolve Fade-In and Fade-Out

Alattar [19] proposed an approach based on the variance of pixel intensities. Fades are detected by recording all negative spikes in the time series of the second difference of the pixel intensity variance, by ensuring that the first order difference of the mean of the video sequence is relatively constant next to the negative spike. Hyeokman et al. [52] showed that different video effects manifest themselves as different patterns on the visual rhythm. Fernando et al. [61] proposed the theoretical fade creation model based on luminosity to detect fade and dissolve. Modification on the above approach is suggested by Truong et al. [62]. Dissolves, fade-in, and fade-out are also discussed in [15,20,63].

Volkmer et al. [64] presented a gradual transition detection approach based on average frame similarity and adaptive thresholds. Albanese et al. [21] proposed an algorithm which use similarity metric based on the animate vision theory. Cai et al. [65] proposed a scheme for identifying the cut and dissolve by using linear perdition for predicting a frame and then prediction error are used to determine existence of short boundaries.

Becós et al. [48] analyzed the existing methods and proposed a unified detection model for gradual transition detection . Han et al. [66] proposed a method which uses techniques of wavelet, FCM clustering , Gaussian weighted Hausdorff distance , similarity of color distribution and motion vector based on 3D wavelet transforms to detect fade, dissolve and wipe. Ling et al. [67] proposed a method which used variance projection function to find distance between the video frames and then a statistical learning method on SVM is used to determine whether the changes of the distance are caused by gradual transition or not. Rong et al. [22] proposed a method which estimate the peaks on the frame to frame difference curve by expectation maximum curve fitting. Each peak contour is approximated by a mixture of Gaussian and uniform distributions. These are used as an input feature for the decision tree classifier to discriminate gradual transition from cuts and motion.

Pedro et al. [68] proposed a fade detection using a non-common feature entropy, a scalar representation of the amount of information conveyed by each video frame. Černekovó et al. [23] have proposed an algorithm for detecting abrupt cuts and fades using the mutual information and the joint entropy between the frames. This method can detect cuts, fade-in and fade-out. Bezerra and Leite et al. [69] used the longest common subsequence between two strings to transform the video slice into 1D signals obtaining a highly simplified representation of the video content. They developed a chain of mathematical morphology operations over these signals to detect cut, fade and wipe.

Liang et al. [70] proposed an algorithm for shot boundary detection which uses mutual information calculated separately for each of the HSV color components. Then a Petri-Net model is describe to detect cut transition and gradual transitions. Jun et al. [71] proposed a method which extracts the color and the edge in different direction from wavelet transition coefficients. Then SVM classifier is used to classify abrupt transition and gradual transition.

Wenzhu and Lihang [72] divided video frames into several different groups through graph-theoretical algorithm, then cut and gradual transitions are detected from the different characteristics between successive frames. Milind and Mukesh [73] proposed a dual stage divide-and-merge approach to detect dissolve transitions. Initially video is divided into non-dissolve frames and dissolve frames, later using histogram comparison with SVD dissolve id detected. Ralph and Bernd [74] proposed an unsupervised approach for gradual transition detection by frame dissimilarity measurements at a lower temporal resolution, then camera motion estimation approach has been used for removal of false alarm. A learning based methodology using a set of features that are specifically designed to capture the differences among hard cuts, gradual transitions and normal sequences at the same time was proposed by Vasileios et al. [75].

2.2 Features Used for Representation of Video Frames

Shot boundary detection algorithm extracts one or more features from a video frame and then different metrics are used to detect shot changes. Thus the large dimensionality of the video is reduced by extracting a small number of features from one or more regions of interest in each video frame [46, 49]. These features are discussed below:

- *Luminance/color*: The average grayscale luminance can be used to characterize a region of interest. Such features are however sensitive to illumination changes. One or more statistics of the values in color space like HSV can be more robust than these simple features [16].
- *Grayscale/color histogram*: Almost all possible variations in calculating intensity or color histogram differences between two continuous frames have been proposed for hard cut detection such as using bin-wise differences, chi-square test, or histogram intersections combined with different color spaces such as RGB, HSV, YIQ, etc. Histograms are insensitive to translation, rotational and zooming camera motion and are easy to compute. Hence it provides a richer feature than simple statistics features based on luminance and is widely used for shot detection [6–8, 10, 11].
- *Image edges*: The edges of objects in the last frame before the hard cut usually cannot be found in the first frame after the hard cut. Zabih et al. exploited this fact and used these features to find edge change ratio .

These features are invariant to illumination changes and several types of motion, but are sensitive to noise and computationally expensive.

- *Transform coefficients based on DCT and DWT*: The transform coefficients in the frequency domain or wavelet domain are related to the spatial domain and can be used as a feature for shot detection [45].
- *Motion*: Motion features are obtained by using the number of motion vectors , distribution of motion vectors and the strength of the residual, derived by block matching method in the optical flow . Motion vectors exhibit relatively continuous changes within a single camera shot, while this continuity will be disrupted between frames across different shots and can be used as a feature for shot detection [76]. But these features are usually coupled with other features and are not useful if there is no motion in the video. The core problem with all motion features arises from the fact that the motion estimation is far more difficult than detecting visual discontinuity, and thus less reliable.

2.3 Major Metrics Used for Shot Boundary Detection

In order to evaluate discontinuity between frames based on the selected features an appropriate metric needs to be chosen. below we provide an overview of major metrics used in SBD.

- *Histogram based metric*: Nagasaka and Tanka [5] compared several simple statistics based on grayscale and color histograms. They found the best results by breaking the image into 16 regions, using a χ^2 test on color histogram of those regions, and discarding the eight largest differences to reduce the effects of object motion and noise. The difference measure is the sum of the absolute bin-wise histogram differences of the consecutive frames. A shot boundary is declared if this difference measure exceeds a threshold.
  Zhang et al. [3] also stated that histogram methods are a good trade-off between accuracy and speed. A 64 bin grayscale histogram is computed over each image. If the histogram difference between consecutive frames exceeds a threshold, a cut is declared. Onur et al. [77] proposed a fuzzy color histogram based shot boundary detection.
- *Statistic based metrics*: Jain et al. [38] computed a likelihood ratio test based on the assumption of uniform second order statistics. A likelihood ratio test is a standard hypothesis test in which a ratio of probabilities is used as the test statistic.

- *Pixel difference based metric*: Pixel difference metrics compare images based on the image intensity map. Nagasaka and Tanka [5] computed a pointwise sum of differences between image pair, whereas Zhang et al. [3] employed a similar measure in which a difference picture is computed and used for shot detection.
- *Edge based metrics*: Zabih et al. [44] proposed a metric that relies on the number of edge pixels that change in a neighboring images. The algorithm is complex as it requires computing edges, registering the images, computing incoming and outgoing edges and computing an edge change ratio. This metric is invariant to illumination changes and is used for flashlight detection in [26, 59, 78].
- *MPEG metrics*: Shot boundary detection in MPEG sequences are very useful due to the attractiveness of processing the compressed data directly. Arman et al. [79] used the coefficients of DCT as a metric to find shot boundaries. Yeo and Liu [45] used DC images for shot boundary detection in MPEG sequences. Each pixel represents the DC value of each 8×8 block in a DC image, which results in significant data reduction. Then a combination of histogram and pixel difference metrics is used for shot detection.

Histogram and statistic based methods are sensitive to lighting changes but are invariant to changes in object motion. The pixel difference comparisons are more robust to lighting changes and are sensitive to motion and camera zooming and panning. The edge based methods are invariant to lighting changes and are generally used in combination with histogram based method and are computation wise very expensive. MPEG metrics are computationally efficient but are sensitive to flashes. A detailed comparison of these metrics for various video data is given in [3, 5–8, 32, 47].

2.4 Decision Methods for Shot Boundary Detection

After applying metrics for shot boundary detection, an algorithm needs to detect where discontinuity exhibit [49]. These boundaries can be found out by applying different thresholds as discussed below:

- Global Threshold: The input to a global thresholding technique could be a time series of feature values of a measure of discontinuity, which in ideal case supposed to show a single peak at hard cut locations. The similarity or dissimilarity of the features computed on consecutive frames is decided by a fixed threshold, which needs to be adjusted for each new video data. Hence it is often impossible to find a single global threshold that works with all kinds of video data.

- Adaptive threshold: Here shot is detected based on the difference of the current feature values from its local neighborhood. Usually a temporal sliding window is centered around the current time instance to represent the local neighborhood. This threshold depends on statistics such as mean and standard deviation of the feature difference metrics within a temporal window [9, 80].
- Probabilistic detection: Here the pattern of specific types of shot transitions are modeled by considering specific probability distributions for the feature difference metrics in each shot and then optimal shot change estimation is performed to detect shot change [10].
- Trained classifier: The shot change is considered as a classification task where frames are classified as shot change or no shot change by using their corresponding features and using a train classifier such as neural network , Support Vector Machine (SVM) and Hidden Markov Model (HMM) [12, 81–85].

2.5 Evaluation Criterion Used in Shot Boundary Detection

The general framework to evaluate and compare various video shot boundary detection algorithms for providing a quality measure has been discussed in detail by Ruiloba et al. [86].

Traditionally Recall and Precision are the two metrics used for evaluation of shot detection algorithms [6, 8, 10, 11, 86].

Recall is defined as

$$R = \frac{C}{C+M} = \frac{C}{D} \tag{2.1}$$

whereas Precision is defined as

$$P = \frac{C}{C+FP} \tag{2.2}$$

where D is the total number of actual shot boundaries (ground truth) in the test video sequence, C is the number of shot boundaries correctly detected by the algorithm, M is the number shot boundaries missed by the algorithm, and FP is the false positives detected by the algorithm.

Also to rank the performance of different algorithms, we used an $F1$ measure [11] which combines recall and precision with equal weight. The

F1 measure is a harmonic mean of Recall and Precision and is given below

$$F1(R, P) = \frac{2 \times R \times P}{R + P} \tag{2.3}$$

2.6 Summary and Conclusions

In this chapter, we provided a survey of papers on the comparison of major shot boundary detection algorithms. Also a detailed survey of papers on wipe detection and shot boundary detection in the presence of flashlight, fire flicker explosion, and motion is provided. We also discussed the feature used for video frame representation and the major metrics used in shot boundary detection. Finally various decision methods and evaluation criterion for shot boundary method were presented.

3

Effective Algorithm for Detecting Various Wipe Patterns

3.1 Introduction

Wipe transition has received less attention of research community, compared to cuts and dissolves, due to the complexity and diversity in its transition patterns. In wipe transitions, the pixels in the current shot are replaced by those in the next shot step by step until the current shot is completely replaced by the next shot as shown in Figures 3.1 and 3.2

This transition involves movement of single or multiple image boundary lines that vary in shapes, moving direction and moving speed. There are more than 30 types of wipe effects commonly used in video editing. These special effect edits are beyond simple shot-to-shot connections, and are often used in news, sports, cartoons, comedy and show programs. Unlike dissolves and fade, it is complicated to model the wipe transition by any single formula as their patterns vary considerably. Hence it is a challenging task to detect the presence of all different types of wipe transitions as well as to identify their individual patterns. Works related to wipe transition detection are given below.

3.2 Related Work and Their Limitations

Alattar [13] developed a model for wipe detection, which exploits the linear change in the mean and variance of frames in the wipe region. The means and the variances of the frames in the wipe region have either a linear or quadratic behavior. Vigorous motion in the scenes composing the wipe sequence causes false alarm as these motion causes gradual change in the means and variance of the frames. Also this algorithm failed in locating the frame range of a wipe transition accurately. Fernando et al. [14] applied the Hough transform to the image boundary to analyze wipe pattern. This method failed when moving

(a)

(b)

Figure 3.1 Video clip from Star War III movie for various wipe type. (a) Horizontal wipe, (b) diagonal wipe

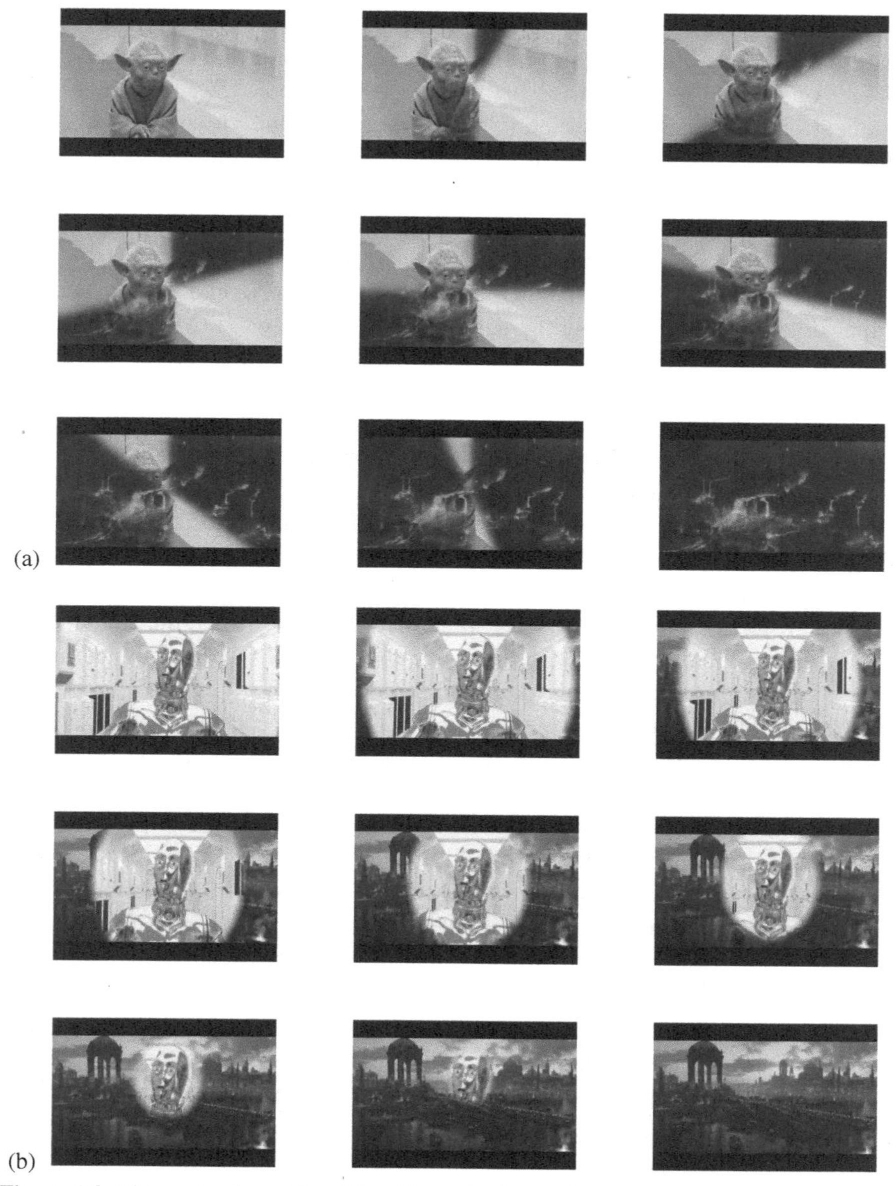

Figure 3.2 Video clip from Star War III movie for various wipe type. (a) Clock wipe, (b) circular wipe

objects are present in the scene in addition to wipes, as these object lines are also detected by the Hough transform and results in false detection. Also this method is able to identify only horizontal, vertical, and diagonal wipes only if one or two wipe boundaries are moving. Kim et al. [52] introduced the concept of visual rhythm, which is an abstraction of the video. They showed that different video effects manifest themselves as different patterns on the visual rhythm. They concluded that all wipes will generate lines from top to bottom in the visual rhythm, and if this line is over a range of frames, instead of vertical line over a single frame, then shot change is due to wipe. Their method could detect only five wipe types mainly horizontal wipe, vertical wipe, diagonal wipe, expanding wipe and absorbing wipe. Major drawback of their method is that camera and object motion could produce the effect of slanted lines in visual rhythm and cause false detection. Also vertical and absorbing wipes show vertical lines on the visual rhythm and are misread as cut. Campisi et al. [54] developed an algorithm based on the trajectory estimation of the boundary line between two successive frames for wipe detection. This method have a limitation of identifying only horizontal, vertical, diagonal, and clock wipe. Also object motion results in false positives due to the same trajectory pattern observed when object moves in one direction. Han et al. [55] used a 3D wavelet transform, Gaussian weighted Hausdorff distance, and the direction of motion vector to identify wipes. This method is limited to horizontal, vertical and diagonal wipe.

Nam et al. [16] used wavelet transform to find direction-emphasized images. These images are projected on a specified angle by Radon transform and then B-Spline interpolation is used to identify wipe region. This scheme is designed to capture vertical, horizontal, diagonal and band wipes with one or two moving boundary lines. This algorithm accurately identifies the frame range of the wipe transition for the designed wipe schemes. Mackowiak et al. [56] employed motion activity, and dominant color descriptors to find wipe region. The algorithm works only for right to left wipe transition. Iwamoto et al. [17] used HSV histogram to find potential wipe candidates, and extract boundary lines from these candidates using the Canny edge detector and the Hough transform. Finally optical flow vectors of the multiple line are calculated to differentiate between object boundary and image boundary. The false positives in their algorithm are caused by moving object lines which the optical flow analysis failed to filter out. Li et al. [18] proposed properties of independence and completeness to characterize ideal wipe. The Bayes rule is applied to each potential wipe to estimate an adaptive threshold for the wipe verification. This method is unable to identify wipes with very fast motion, since the motion estimation approach used is not able to track such motions. Limitations of the existing methods are summarized below:

- Limited to detection of primary pattern, like vertical, horizontal, and diagonal wipes with only one or two moving lines.
- Unable to detect nonlinear and nonrigid changing patterns and boundary.
- Difficult to distinguish wipe from camera and object motion, which causes miss detection and false alarm.
- Deficient in locating the frame range of a wipe transition accurately.
- Lines formed by moving object misdetected as image boundary lines.
- Detection of motion vector and optical flow for each frame is computationally expensive.
- The thickness of the wipe border has an effect on the detection performance as some wipe transitions may have a blurred wipe border and may not be effectively picked up.
- Difficult to detect blur border if an exiting and entering scene both have dark background.

The rest of this chapter is organized as follows. In Section 3.3 various wipe types observed in the test video sequences is described. Section 3.4 gives a detailed description of the proposed method. Experimental results using the proposed algorithm are discussed in Section 3.5. Performance comparison with the other wipe transition detection methods are shown in Section 3.6 in terms of Recall, Precision, F1 measure and Detection Rate. Demonstration of various patterns obtained using the proposed algorithm is described in Section 3.7.

3.3 Various Wipe Types Observed in the Test Video Sequence

3.3.1 Test Video Sequence

We have considered movies such as Star Wars III, Star Wars I, Jodhaa Akbar, BhoothNath, and Tomcat for the test video sequence. These movies are manually observed frame by frame for ground truth to find the desired detection. we found 31 different wipe patterns and are broadly categorized into horizontal wipe, (denoted by HWP1-HWP7) diagonal wipe, (denoted by DWP1-DWP6) vertical wipe, (denoted by VWP1-VWP7) and special wipe (denoted by SWP1-SWP4) as shown in Figures 3.3 (23 pattern which are denoted by HWP1-HWP7, VWP1-VWP7, SWP1-SWP4) and 3.4 (8 pattern which are denoted by Tom1-Tom8). Out of these 31 wipe types HWP1, HWP2, VWP1-VWP4, SWP1-SWP4, Tom1-Tom8 are complex wipe. These are categorized according to the number of moving wipe lines, their moving direction and shapes.

These figures are actual frames from the above mentioned movies. These wipes vary in border shapes, moving direction, moving speed, and total wipe span. Number of wipe sequences available in SWIII, SWI, JA and BN movies with different patterns and information about number of frames with wipe and wipe range is shown in Table 3.1.

To test the robustness of the algorithm, we consider the frames from the video where camera and object motion are present before or after wipe . In Star Wars III movie 959 frames with wipes were found followed by 1086 frames , 342 frames , and 109 frames in Star Wars I, Jodha Akbar, and Bhooth Nath movie respectively.

We also consider another movie Tomcat for the test video sequence which consists of complex scenes. We found 13 different wipe patterns in this movie, out of which 8 wipe patterns were not observed in the previous 4 tested movie as shown in Figure 3.4.

The number of wipe sequences available in Tomcat movie with different patterns and information about the number of frames with wipe and wipe range is shown in Table 3.2.

3.4 Proposed Algorithm for Wipe Transition Detection

Wipe transition involves movement of single or multiple image boundary lines. These geometrical features can be used to detect wipes by the Hough transform. But image boundaries are difficult to detect if wipe border is blurred, and exiting and entering scene have dark background. Also it is very difficult to differentiate between lines formed by object boundary and image boundary. Wipe transition is difficult to detect due to complexity and diversity in its transition patterns.

In this chapter, we propose an effective algorithm for wipe transition detection, as reported in [87]. In our proposed algorithm, the pixels responsible for scene change are detected by observing intensity change during wipe for every pixel and then adaptive threshold is used to convert these pixels into a binary image. Wavelet transform and morphological operations are then applied to this binary image to find a moving strip. This process removes most of the object boundary lines and retains wipe boundaries, and then the Hough transform is applied to detect and categorize various wipe types.

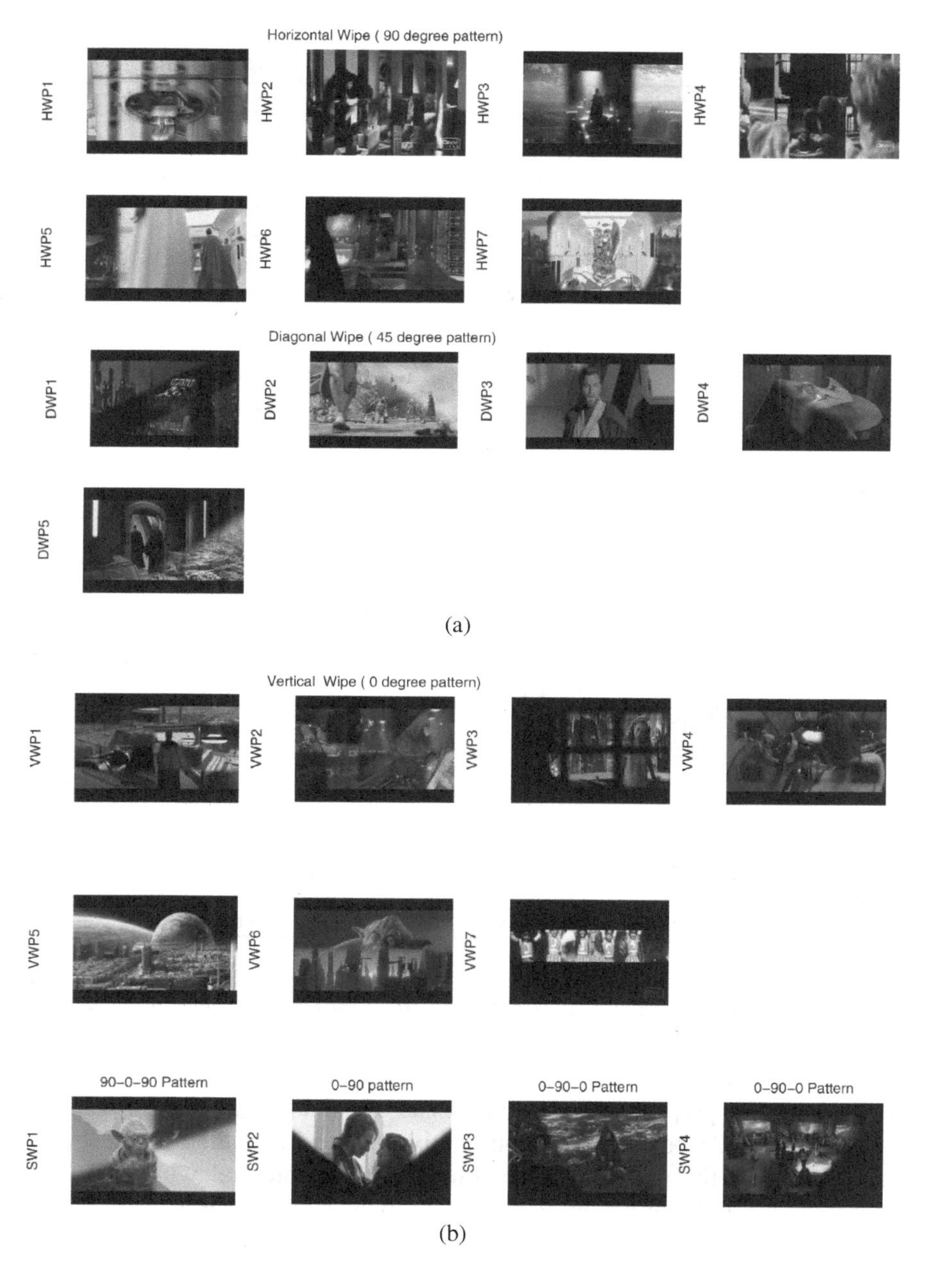

Figure 3.3 Various wipe patterns observed in the test video sequence. (a) Various horizontal and diagonal wipe types, (b) various vertical and special wipe types

Table 3.1 Wipe transition types in each movie

Wipe Type	Video→	SWIII	SWI	JA	BN
Horizontal Wipe	HWP1	2	–	–	–
	HWP2	–	–	–	2
	HWP3	3	–	1	–
	HWP4	–	–	–	3
	HWP5	2	10	3	–
	HWP6	2	6	5	–
	HWP7	5	–	–	–
	Total HW	14	16	9	5
Diagonal Wipe	DWP1	1	–	–	–
	DWP2	1	–	–	–
	DWP3	3	1	–	–
	DWP4	2	2	–	–
	DWP5	2	9	–	–
	Total DW	9	12	–	–
Vertical Wipe	VWP1	1	–	–	–
	VWP2	1	–	–	–
	VWP3	1	–	–	–
	VWP4	1	–	–	–
	VWP5	2	5	–	–
	VWP6	1	8	1	–
	VWP7	–	–	–	2
	Total VW	7	13	1	2
Special Wipe	SWP1	1	–	–	–
	SWP2	1	–	–	–
	SWP3	1	–	–	–
	SWP4	1	–	–	–
	Total SW	4	–	–	–
Total wipe types	Movie wise	34	41	10	7
Total wipe frames	Movie wise	959	1086	342	109
Frame range	Min:Max	12–32	20–33	11–58	8–22

Applying the proposed algorithm to the whole video to detect wipes is computationally expensive. In order to decrease the computational load of the proposed algorithm, we propose a preprocessing step as a first stage of the algorithm. The preprocessing step consists of calculating mean of statistical image difference between the consecutive frames, and then a threshold is used to obtain the potential wipe frames. These potential wipe frames are processed through our proposed algorithm to detect and identify various wipe types.

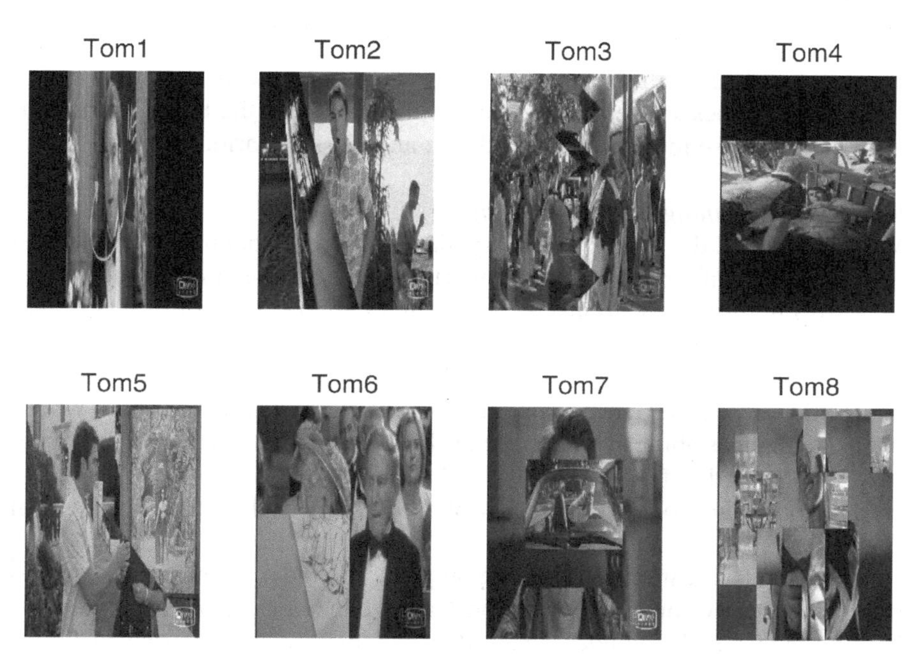

Figure 3.4 Various new wipe types observed in Tomcat movie which were not present in previous four videos

Table 3.2 Wipe transition types in Tomcat movie

Wipe Pattern	Wipe type	Tomcat
Horizontal Wipe	Tom1	10
	Tom2	01
	Tom3	01
	HWP5	02
	HWP6	01
	HWP7	04
Diagonal Wipe	DWP4	01
Vertical Wipe	Tom4	01
	VWP6	01
Special Wipe	Tom5	02
	Tom6	02
	Tom7	03
	Tom8	01
Total wipe types	Tomcat Movie	30
Total Wipe Frames	Tomcat Movie	673
Frame Range	Min:Max	13–35

3.4.1 Proposed Algorithm without Preprocessing

The proposed algorithm consists of two stages. In the first stage moving strips due to wipes are determined, and in the next stage the Hough transform is applied to these moving strips to detect and classify various wipe types.

Stage I: Determining the moving strip in wipe region
Moving strip is the image boundary line formed on boundary of the two image regions of an exiting and entering scene in the transitional frame. This strip is formed by connected edges and they move in a time sequence. According to Li et al. [18], in an ideal wipe transition, every pixel changes its value only once during wipe and all pixels will change their values after completion of wipes. Hence to find the moving strip, we find the pixels whose intensity changes abruptly due to scene change. Detail Steps to obtain moving strip is described in [87].

Demonstration of several stages of the proposed algorithm is shown in Figure 3.5.

The above mentioned steps are applied to find moving edges due to wipe instead of Canny edge detector. Canny edge detector detects all edges in the scene which includes edges due to wipes as well as due to object boundaries. Whereas by using our approach, most of the edges due to the object boundaries are eliminated and only true wipe edges are retained as shown in Figures 3.6(a) and 3.6(b). The thickness of the wipe border has an effect on the detection performance as some wipe transitions may have blurred wipe borders and may not be effectively picked up, and also blur borders are difficult to detect if exiting and entering scenes both have dark backgrounds. Figures 3.6(c) and 3.6(d) show wipe transitions with blur borders, and the results of our proposed approach clearly indicate the advantage over Canny edge detector in finding moving strips in such cases. We used the approach followed by Iwamoto et al. [17] for comparison, where they first find pixelwise differences between consecutive frames and then Canny edge detector is applied to find moving strips. Our proposed method used to detect moving strips, takes care of both the above discussed limitations.

Stage II: Applying the Hough transform to detect and classify wipe types
The basic Hough transform, also called the standard Hough transform, has established itself as a default technique for extracting global features such as straight line, circle etc. from an image. This algorithm uses an array called

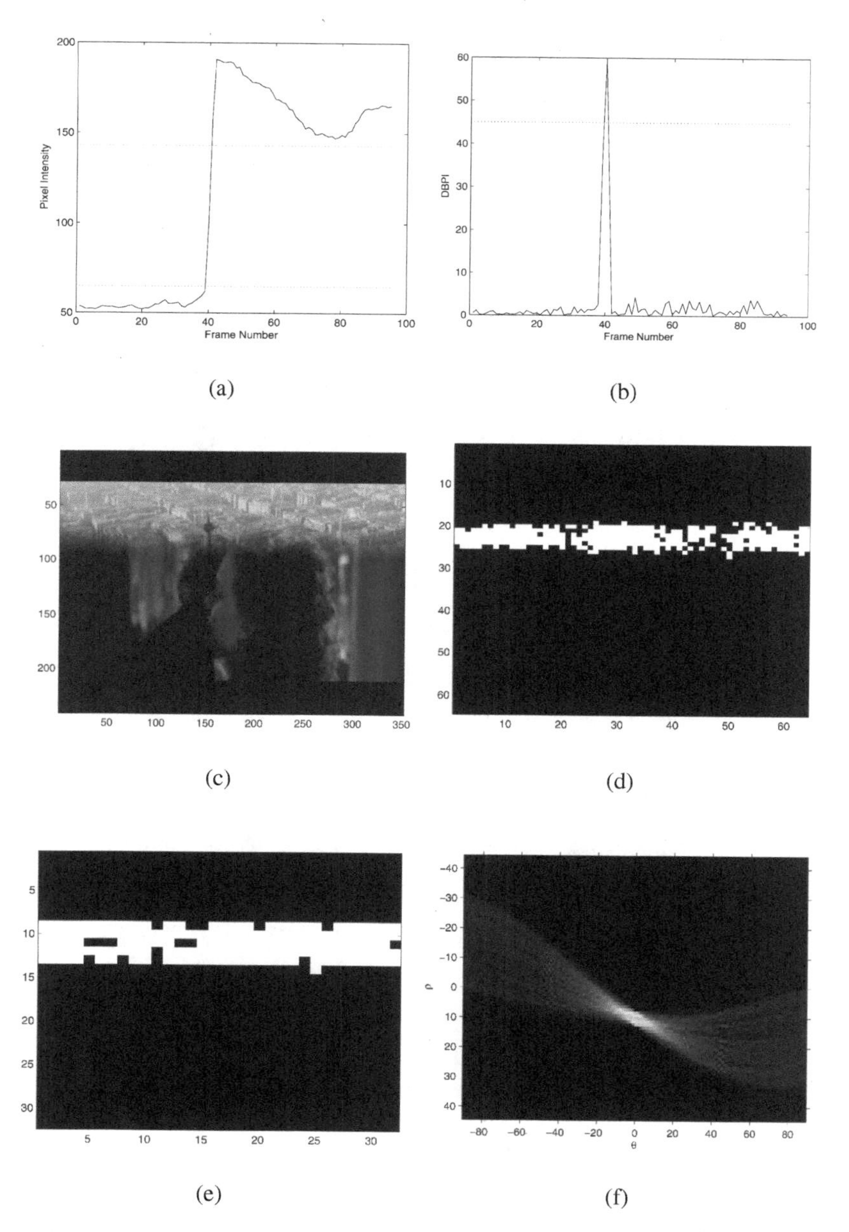

Figure 3.5 Demonstration of several stages of the proposed algorithm: (a) pixel intensity, (b) difference between pixel intensity, (c) frame from Star War III movie, (d) binary image, (e) moving strip, (f) highest voted angle after applying the Hough transform

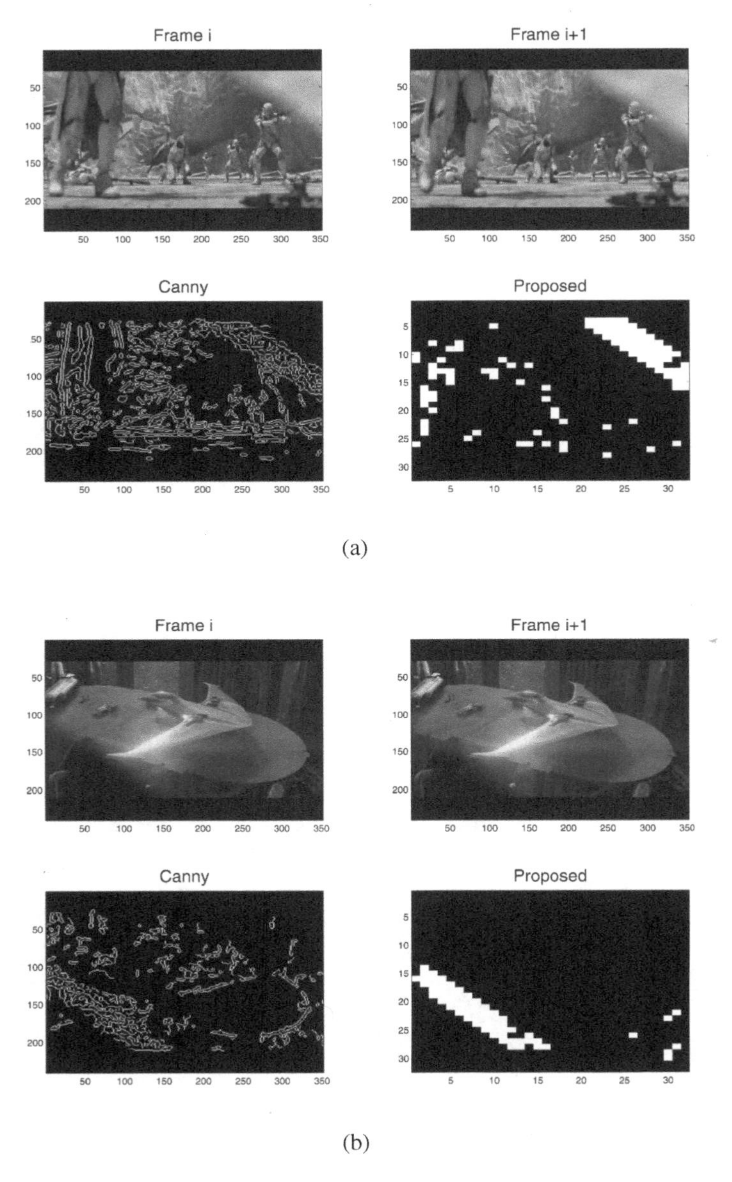

Figure 3.6 Comparison of the proposed approach over Canny edge detector in wipe boundary detection. (a) Comparison when edges due to object are present in addition to wipe boundary edges. (b) Comparison when edges due to object are present in addition to blur wipe boundary edges

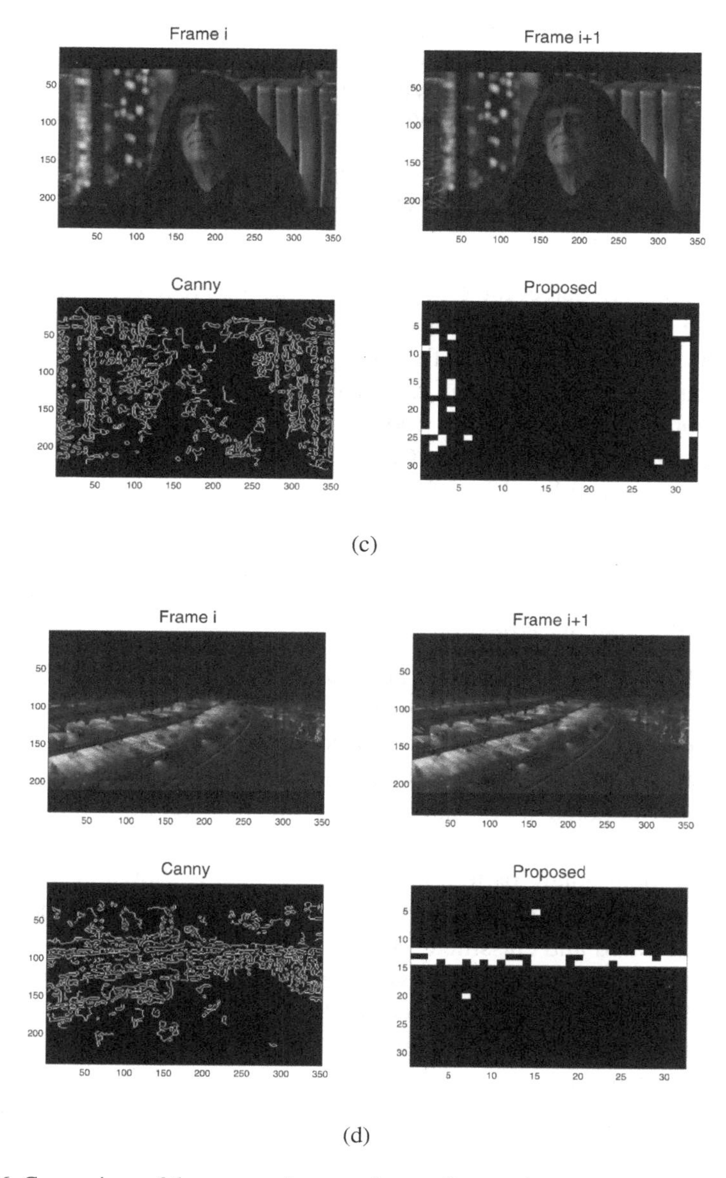

Figure 3.6 Comparison of the proposed approach over Canny edge detector in wipe boundary detection. (Continued) (c) Comparison when wipe border is blurred and exiting and entering scene have dark background. (d) Comparison when wipe border is blurred, exiting and entering scene have dark background, and edges due to object and wipe border overlap

Table 3.3 Various wipe types and corresponding gradient pattern observed

Wipe notation	Gradient pattern
HWP1 to HWP7	90 pattern
VWP1 to VWP7	0 pattern
DWP1 to DWP5	45 pattern
SWP1	90-0-90 pattern
SWP2	0-90 pattern
SWP3 and SWP4	0-90-0 pattern

parameter space to detect the existence of a line $y = a \times x + b$ for each edge pixel in the edge image. The Hough transform algorithm determines if there is enough evidence of a line at that pixel. It calculates the parameter values and accumulates the cells in the parameter space according to all the pixels. An equation expressed in polar coordinates $\rho = x \times \cos\theta + y \times \sin\theta$ is used instead of an equation in the Cartesian coordinate system, which cannot represent the vertical line [88, 89].

We obtain the gradient pattern plot from the moving strip obtained in stage I using the steps described in [87].

1. Detection and classification of wipes.
    - If the gradient pattern is constant at 90° (as shown in Figure 3.7(a) and denoted as 90 pattern) for more than 8 frames, then wipe is declared and classified as horizontal wipe.
    - If the gradient pattern is constant at 0° (as shown in Figure 3.7(b) and denoted as 0 pattern) for more than 8 frames, then wipe is declared and classified as vertical wipe.
    - If the gradient pattern is constant at 45° (as shown in Figure 3.7(c) and denoted as 45 pattern) for more than 8 frames, then wipe is declared and classified as diagonal wipe.
    - If the gradient pattern is varying from 90-0-90 degrees (as shown in Figure 3.7(d) and denoted as 90-0-90 variation pattern) or 0-90-0 degree (as shown in Figure 3.7(e) and denoted as 0-90-0 variation pattern) or 0-90 degree (as shown in Figure 3.7(f) and denoted as 0-90 variation pattern) for more than 8 frames, wipe is declared and classified as special wipes.

Table 3.3 gives details about the gradient pattern observed for each wipe type.

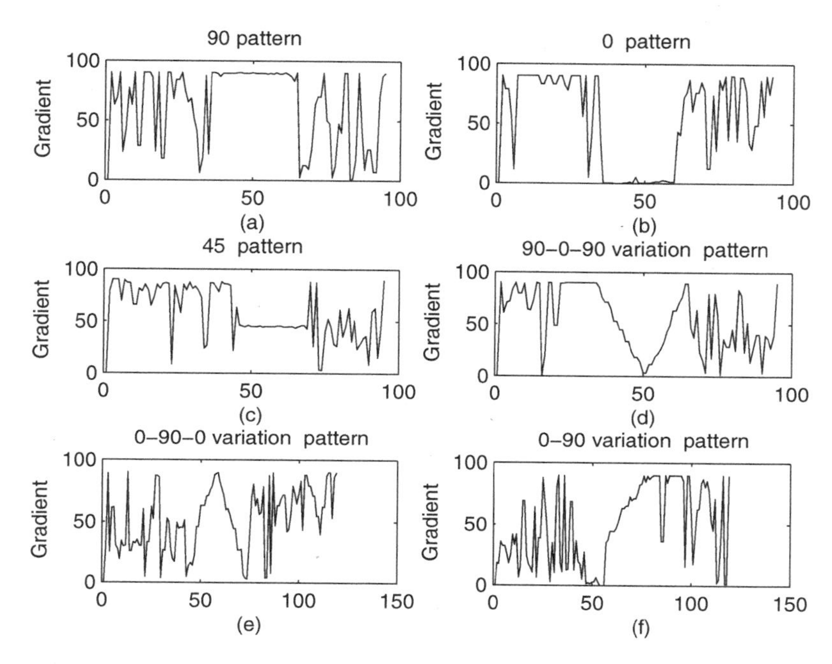

Figure 3.7 Various gradient pattern observed for different wipes types: (a) 90 degree pattern observed for horizontal wipe (b) 0 degree pattern observed for vertical wipe (c) 45 degree pattern observed for diagonal wipe (d) 90-0-90 degree variation pattern observed for special wipe (SWP1) (e) 0-90-0 degree variation pattern observed for special wipe (SWP3, SWP4) (f) 0-90 degree variation pattern observed for special wipe (SWP2)

3.4.2 Proposed Algorithm with Preprocessing

Applying the Hough transform on all the frames considered for analysis is computationally expensive. Hence we propose a preprocessing method which significantly reduces this computational load of the algorithm. In this method, first the statistical image frame difference, denoted as *MOSID*, between the consecutive frames is calculated, and then the threshold denoted as S_T, is used to find the potential wipe frames. Steps for finding the potential wipe frames are described in [87].

Then these potential wipe frames are processed by *Stage-I* and *Stage-II* as discussed in Section 3.4.1 to detect and classify various wipe types.

3.5 Experimental Results and Discussion

The proposed algorithm with and without preprocessing stage have been tested on various movies, such as Star Wars III, Star Wars I, Jodhaa Akbar, BhoothNath, and Tomcat on the test video sequences shown in Tables 3.1 and 3.2.

3.5.1 Evaluation Criterion

We used Recall, Precision and F1 measure as discussed in Section 2.5 as a evaluation metric to compare shot boundary detection algorithm. However wipe transition takes place over a certain range of frames unlike abrupt transition which occurs at a single frame. Hence it is not possible to evaluate if the methods could actually identify different types of wipes by above metric. Therefore we need an additional evaluation metric which can elaborate details about how many wipe types are correctly detected by the algorithm [16]. This metric is defined as Detection Rate () and is given by

$$\text{Detection Rate (DR)} = \frac{\text{Number of wipe types correctly detected (Hit)}}{\text{Number of actual wipe types in the video (Actual)}} \tag{3.1}$$

The range of the wipe varies from 8 frames to 58 frames in the test video sequence as shown in Table 3.1. For the full range of the one wipe frames, one wipe type is considered and the number of actual wipe types are found out and shown in Table 3.1 (total wipe types movie wise). Any partial detection is considered as correct detection (Hit) if more than 10 percent of its wipe range is correctly declared [16].

3.5.2 Experimental Results Using the Proposed Algorithm

Frame based R, P, and F1 measure performance comparison between the proposed method with and without preprocessing step is shown in Table 3.4.

Results obtained with the proposed algorithm with preprocessing step as shown in Table 3.4, by considering tuning parameter $\alpha = 0.125$ and $\beta = 0.125$ (selection of the tuning parameters and its effect on Recall, Precision, and F1 measure is discussed in detail in Section 3.5.3). The proposed algorithm is able to detect 28 wipe types out of 31 wipe types found in this test video sequence and results in high Recall and higher Detection Rate. Our algorithm failed to detect wipe types VWP3 and VWP4 due to its nonlinear and nonrigid changing patterns and boundary. Our proposed algorithm with

Table 3.4 Frame based R, P, and F1 measure performance comparison between the proposed algorithm without and with preprocessing

Algorithm	Video→	SWIII	SWI	JA	BN
Proposed algorithm without preprocessing	R	90.72	91.53	98.54	88.07
	P	95.31	98.91	91.71	87.73
	F1	92.95	95.07	95.00	87.89
Proposed algorithm with preprocessing	R	89.36	89.77	96.78	84.40
	P	94.27	98.18	91.43	87.61
	F1	91.74	93.78	94.02	85.97

Table 3.5 Frame based R, P, and F1 measure performance for Tomcat movie using proposed algorithms

Algorithm	Video→	Tomcat
Proposed algorithm without preprocessing	R	94.50
	P	95.21
	F1	94.85
Proposed algorithm with preprocessing	R	93.02
	P	93.29
	F1	93.15

and without preprocessing step is also tested on Tomcat movie and the results are shown in Table 3.5.

We found 13 different wipe patterns in this movie, out of which 8 wipe patterns were not observed in the previous 4 tested movies as shown in Figure 3.4.

Wipe type Tom1 and wipe type Tom4 wipes are different from wipe type HWP3 and wipe type VWP7 wipes. In Tom1 and Tom4, the current scene is replaced by a black scene using wipe transition and then a black scene is replaced by next scene using wipe transition, whereas in HWP3 and VWP7 the current shot is replaced by next shot using wipe transition.

3.5.3 Selection of the Tuning Parameters and Their Effect on Recall and Precision

We extensively varied the values of the tuning parameter (α and β) in preprocessing step stated in [87] and tested on various movie videos for Recall and Precision. The effect on the performance due to the variation of these parameters has been investigated and the results are reported in Table 3.6 for Jodhaa Akbar movie. We observed the same trend for other movies too and is demonstrated in Tables 3.7, 3.8, and 3.9 for movie Tomcat, BhoothNath, and Star War III, respectively. Based on the extensive simulation results for

Table 3.6 Tuning parameters and its effect on R, P, and F1 for Jodhaa Akbar movie

TP		Results on Jodhaa Akbar movie		
α	β	R	P	F1
0.0625	0.0625	98.54	89.63	93.87
0.125	0.125	96.78	91.43	94.02
0.25	0.25	88.30	98.05	92.91

Table 3.7 Tuning parameters and its effect on R, P, and F1 for Tomcat movie

TP		Results on Tomcat movie		
α	β	R	P	F1
0.0625	0.0625	93.61	91.70	92.64
0.125	0.125	93.02	93.29	93.15
0.25	0.25	81.28	95.13	87.66

various tuning parameters as shown in Tables 3.6, 3.7, 3.8, and 3.9, it is empirically observed that the choice of $\alpha = 0.125$ and $\beta = 0.125$ provide a reasonable trade off between Recall and Precision. It has been clearly observed that the lower values of the tuning parameter results in the drop of Precision value and higher values cause degradation in Recall value. Our exhaustive investigation results also effectively demonstrate that the choice of $\alpha = 0.125$ and $\beta = 0.125$ provide the best value for F1 measure, whereas lower or upper values of tuning parameter degrades the value of F1 measure and is the main reason for choosing these values.

3.6 Performance Comparison with Other Wipe Transition Detection Methods

Performance comparison of the proposed algorithm, with and without preprocessing, has been compared with Alattar [13], Fernando et al. [14], and Nam et al. [16], for the same test video sequences shown in Tables 3.1 and 3.2. We selected these algorithms for comparison, since most of the successful methods suggested for wipe detection [16], including the recent algorithms

Table 3.8 Tuning parameters and its effect on R, P, and F1 for BhoothNath movie

TP		Results on BhoothNath movie		
α	β	R	P	F1
0.0625	0.0625	88.07	82.05	84.95
0.125	0.125	84.40	87.61	85.97
0.25	0.25	77.06	92.31	83.99

Table 3.9 Tuning parameters and its effect on R, P, and F1 for Star War III movie

TP		Results on Star War III movie		
α	β	R	P	F1
0.0625	0.0625	89.78	93.59	91.64
0.125	0.125	89.36	94.27	91.74
0.25	0.25	85.30	95.23	89.99

[17, 18] have been compared with these algorithms. Above algorithms have been tested for four movie videos, Star Wars III (SWIII), Star Wars I (SWI), Jodhaa Akbar (JA), and Bhooth Nath (BN).

3.6.1 Performance Comparison

Frame based R, P, and F1 measure performance comparison between the proposed method with and without preprocessing step, and the other tested algorithms are shown in Table 3.10. The performance comparison based on detection of wipes (Detection Rate) is shown in Table 3.11. Results are obtained with the proposed algorithm with preprocessing step as shown in Table 3.10, by considering tuning parameter $\alpha = 0.125$ and $\beta = 0.125$ (selection of the tuning parameters and its effect on Recall, Precision, and F1 measure is discussed in detail in Section 3.5.3).

Our proposed algorithm without the preprocessing step is a general method and is independent of the tuning parameters, but requires more computations. Whereas the proposed algorithm with the preprocessing step depends on the tuning parameters, but requires less computation with a slight degradation in F1 score.

The algorithm by Alattar [13] is based on statistical features and gives false detection for large camera and object motion. As the method is based on statistical features, it detects most of the wipe types but fails in accurately locating the frame range of wipe and results in low Recall but with comparable Detection Rate. As this method does not follow some common pattern, it is difficult to say whether it method will fail or succeed in detecting a particular wipe type. This algorithm is also unable to differentiate between different wipe types like horizontal, vertical, etc.

Fernando et al. [14] used mean square of the image frame difference to find wipe strip. If moving objects are present in the scene in addition to wipe boundary, then binary image shows edges due to wipe as well as edges due to object. Angle of the edge due to wipe and object may be different, and if the Hough transform is applied to such frames to find average gradient, it may

Table 3.10 Frame based R, P, and F1 measure performance comparison between the proposed algorithm and other tested algorithms

Algorithm	Video→	SWIII	SWI	JA	BN
Proposed algorithm without preprocessing	R	90.72	91.53	98.54	88.07
	P	95.31	98.91	91.71	87.73
	F1	92.95	95.07	95.00	87.89
Proposed algorithm with preprocessing	R	89.36	89.77	96.78	84.40
	P	94.27	98.18	91.43	87.61
	F1	91.74	93.78	94.02	85.97
Alattar [13]	R	65.69	73.57	79.23	74.31
	P	70.31	79.03	86.03	48.21
	F1	67.92	76.20	82.49	58.48
Fernando et al. [14]	R	46.19	50.09	66.37	23.85
	P	84.70	81.55	88.32	43.33
	F1	59.77	62.06	75.78	30.76
Nam et al. [16]	R	67.67	84.53	88.88	55.04
	P	95.44	94.93	92.12	81.08
	F1	79.19	89.42	90.47	65.56

Table 3.11 Performance comparison based on detection of wipes

Algorithm	Video→	SWIII	SWI	JA	BN
Proposed algorithms	Actual	34	41	10	07
	Hit	32	41	10	07
	Miss	02	00	00	00
	DR	94.12	100	100	100
Alattar [13]	Actual	34	41	10	07
	Hit	30	38	09	06
	Miss	04	03	01	01
	DR	88.24	92.68	90	83.33
Fernando et al. [14]	Actual	34	41	10	07
	Hit	24	30	09	03
	Miss	10	11	01	04
	DR	70.59	73.17	90	42.86
Nam et al. [16]	Actual	34	41	10	07
	Hit	29	39	08	04
	Miss	05	02	02	03
	DR	85.29	95.12	80	57.14

not give constant angle pattern and results in miss detection. If moving objects are present before or after wipe for long duration, then average gradient may give constant angle pattern and results in false positives. If wipe changes its direction or involves more than two lines, then average gradient pattern may not be constant after applying the Hough transform. This algorithm mainly failed to detect following wipe types: HWP1, HWP7, VWP2, VWP3, VWP4, SWP1, SWP2, SWP3, and SWP4 and results in low Precision and lower Detection Rate.

In the algorithm proposed by Nam et al. [16], one of the condition for wipe is that, the wipe transition length must lie between predefined minimum length 15 to maximum length 24. For a reasonable trade off between Recall and Precision, we slightly modified this condition by minimum length 8 to maximum length 58 as per our test video data. This algorithm is unable to detect following wipe types: HWP1, HWP2, VWP1, VWP2, VWP3, and VWP4. This is the main reason for low Recall and comparatively lower Detection Rate in this algorithm. Though this algorithm is not designed and tested by Nam et al. [16] for the HWP7, SWP1, SWP2, SWP3, and SWP4 wipe types, we find that this method is able to detect such wipe types due to its linear like transition behavior and straight line boundaries. This algorithm is unable to locate frame range accurately for this wipe types, but it accurately detects the frame range of other detected wipes and results in higher precision.

The proposed algorithm is able to detect 28 wipe types out of 31 wipe types found in this test video sequence and results in high Recall and higher Detection Rate. Our algorithm failed to detect wipe types VWP3 and VWP4 due to its nonlinear and nonrigid changing patterns and boundary. The Hough transform failed to find differentiable gradient pattern for these wipe types. Our algorithm is able to find out the wipe range accurately in most of the wipe types excluding HWP7 and diagonal wipes, where scene change region is very small during starting and end of the wipe. This is the main reason for high precision in the proposed algorithm. The proposed algorithm also discriminates wipes from object and camera motion and is demonstrated in Section 3.7.1. Overall our proposed algorithm gives better tradeoff between Recall and Precision as compared to other compared algorithms and detects most of the wipe types and their range accurately. Our proposed algorithm successfully differentiates between horizontal, vertical, diagonal and special wipe types by observing gradient patterns satisfied by them.

The horizontal, diagonal, and vertical wipes mentioned in Table 3.2 satisfy 90, 45 and 0 patterns, respectively. For type Tom5 we have observed a new 90-0-90-0-90 pattern, whereas for wipe type Tom6 and wipe type Tom7 we observed pattern of 0 and 90 degree alternately. Our proposed method failed to observe any pattern for wipe type Tom8 because it consists of a very complex wipe structure created by digital effects.

Table 3.12 Computational time for the different detection methods

Algo.	Ref. [13]	Ref. [14]	Proposed	Ref. [16]
CT	19.34	97.22	116.85	148.95

3.6.2 Computational Time for the Different Detection Methods

All algorithms were implemented in Matlab 7.0 and the average computational costs were obtained by running the program on Intel P4, CPU 2.40 GHz with 512 MB RAM. Table 3.12 shows the computational time for the different detection methods, where the computational time (denoted as CT) is in milliseconds per frame.

3.7 Demonstration of Various Patterns Obtained Using the Proposed Algorithm

3.7.1 Discriminating Wipe from Object Motion and Demonstration of Horizontal Wipe

To show the effectiveness of the proposed algorithm for discriminating wipe from object motion, 95 frames were considered from Star Wars III movie. Ground truth for the above clip is wipe transition from 46 to 58 frames, followed by fast camera as well as object motion. Figure 3.8(a) shows the frames from this video sequence. The output of the proposed algorithm is shown in Figure 3.8(b). A 90 degree pattern is clearly observed for frame 46 to 58 and declared as horizontal wipe, whereas potential frames 59 to 90 due to camera and object motion do not follow these patterns. The results clearly indicate that using the proposed algorithm camera and object motion can be clearly distinguished.

3.7.2 Demonstration of Diagonal Wipe

To demonstrate the results of diagonal wipe, we considered 95 frames from the movie Star War I. Ground truth for the above clip is wipe transition from 44 to 71 frames. Figure 3.9(a) shows the frames from this video sequence. The proposed algorithm is applied to this video clip and result is shown in Figure 3.9(b). A 45 degree pattern is clearly observed for frame 44 to 71 and declared as a diagonal wipe.

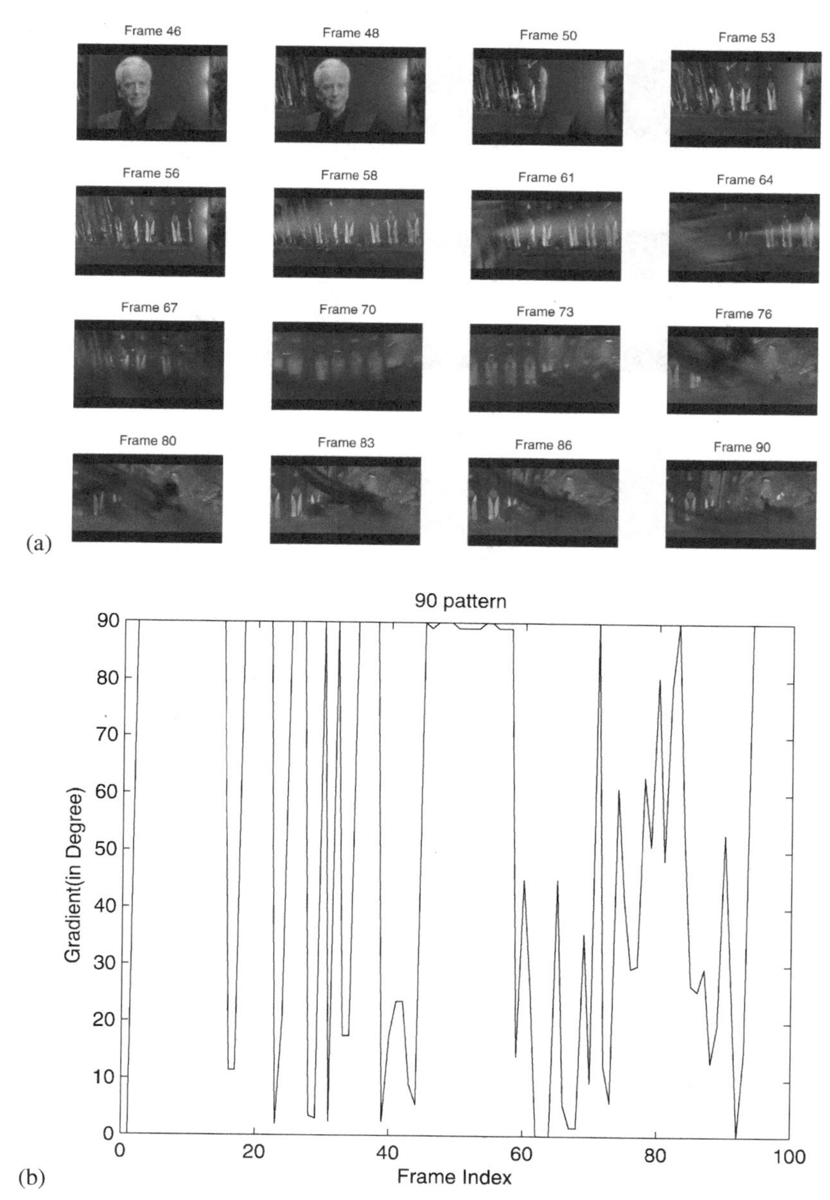

Figure 3.8 (a) Video clip from Star War III movie for horizontal wipe. (b) Gradient pattern satisfied by it

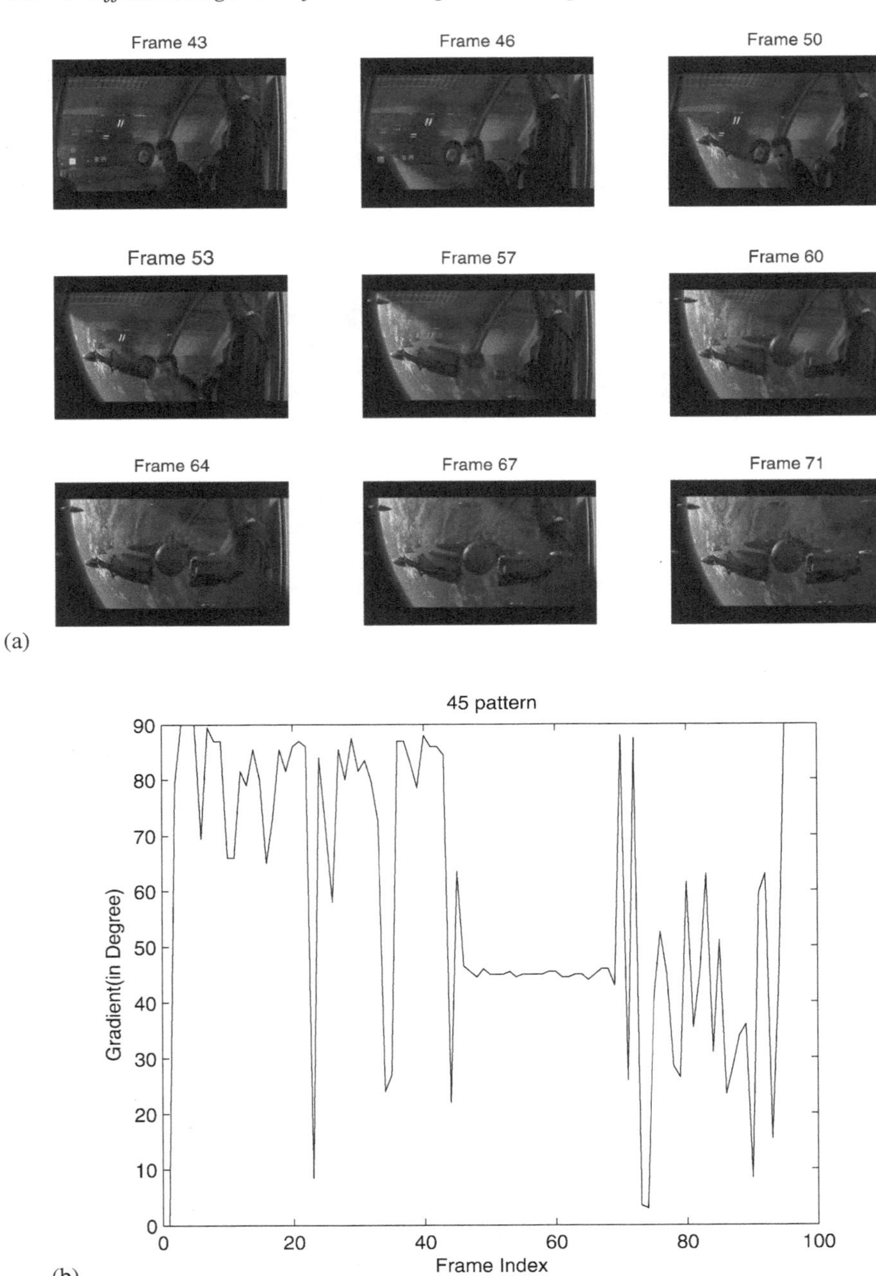

Figure 3.9 (a) Video clip from Star War I movie for diagonal wipe. (b) Gradient pattern satisfied by it

3.7.3 Demonstration of Vertical Wipe

To demonstrate the results of vertical wipe, we considered 93 frames from the movie Star War I. Ground truth for the above clip is wipe transition from 36 to 60 frames. Figure 3.10(a) shows the frames from this video sequence. The proposed algorithm is applied to this video clip and result is shown in Figure 3.10(b). A 0 degree pattern is clearly observed for frame 36 to 60 and declared as a vertical wipe.

3.7.4 Demonstration of Special Wipe SWP1

To demonstrate the results of this special wipe, we considered 95 frames from the movie Star War I. Ground truth for the above clip is wipe transition from 35 to 67 frames. Figure 3.11(a) shows the frames from this video sequence. The proposed algorithm is applied to this video clip and result is shown in Figure 3.11(b). A 90-0-90 degree pattern is clearly observed for frame 36 to 60 and declared as a vertical wipe.

3.7.5 Demonstration of Special Wipe SWP3 and SWP4

To demonstrate the results of this special wipe, we considered 119 frames from the movie Star War III. Ground truth for the above clip is wipe transition from 44 to 72 frames. Figure 3.12(a) shows the frames from this video sequence. The proposed algorithm is applied to this video clip and the result is shown in Figure 3.12(b). A 0-90-0 degree pattern is clearly observed for frame 36 to 60 and declared as a vertical wipe.

3.7.6 Demonstration of Special Wipe SWP2

To demonstrate the results of this special wipe, we considered 119 frames from the movie Star War III. Ground truth for the above clip is wipe transition from 48 to 80 frames. Figure 3.13(a) shows the frames from this video sequence. The proposed algorithm is applied to this video clip and the result is shown in Figure 3.13(b). A 0-90-0 degree pattern is clearly observed for frame 36 to 60 and declared as a vertical wipe.

3.8 Summary and Conclusions

In this chapter, we propose an algorithm for wipe transition detection. In the proposed algorithm, first the moving strip due to wipe is obtained, and then

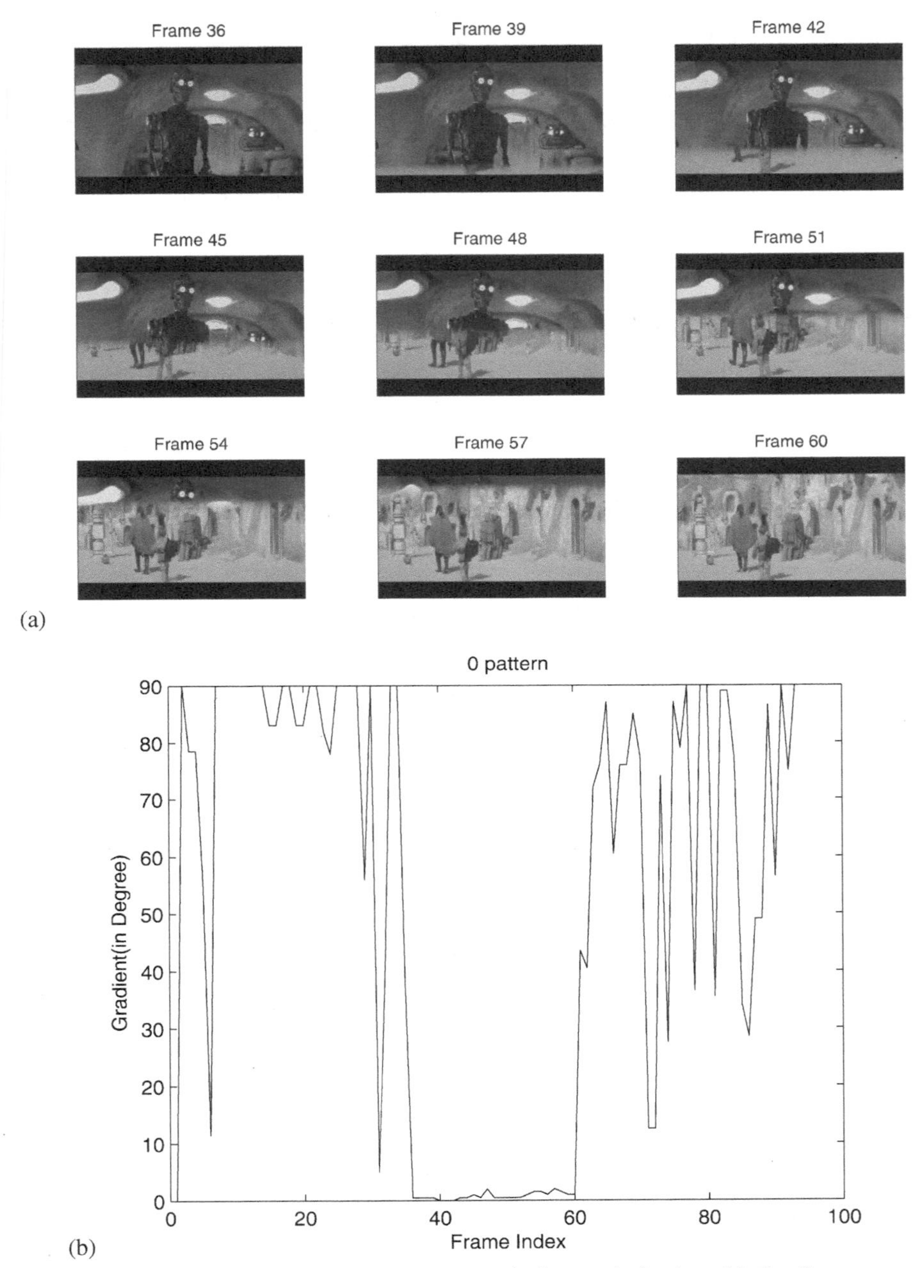

Figure 3.10 (a) Video clip from Star War I movie for vertical wipe. (b) Gradient pattern satisfied by it

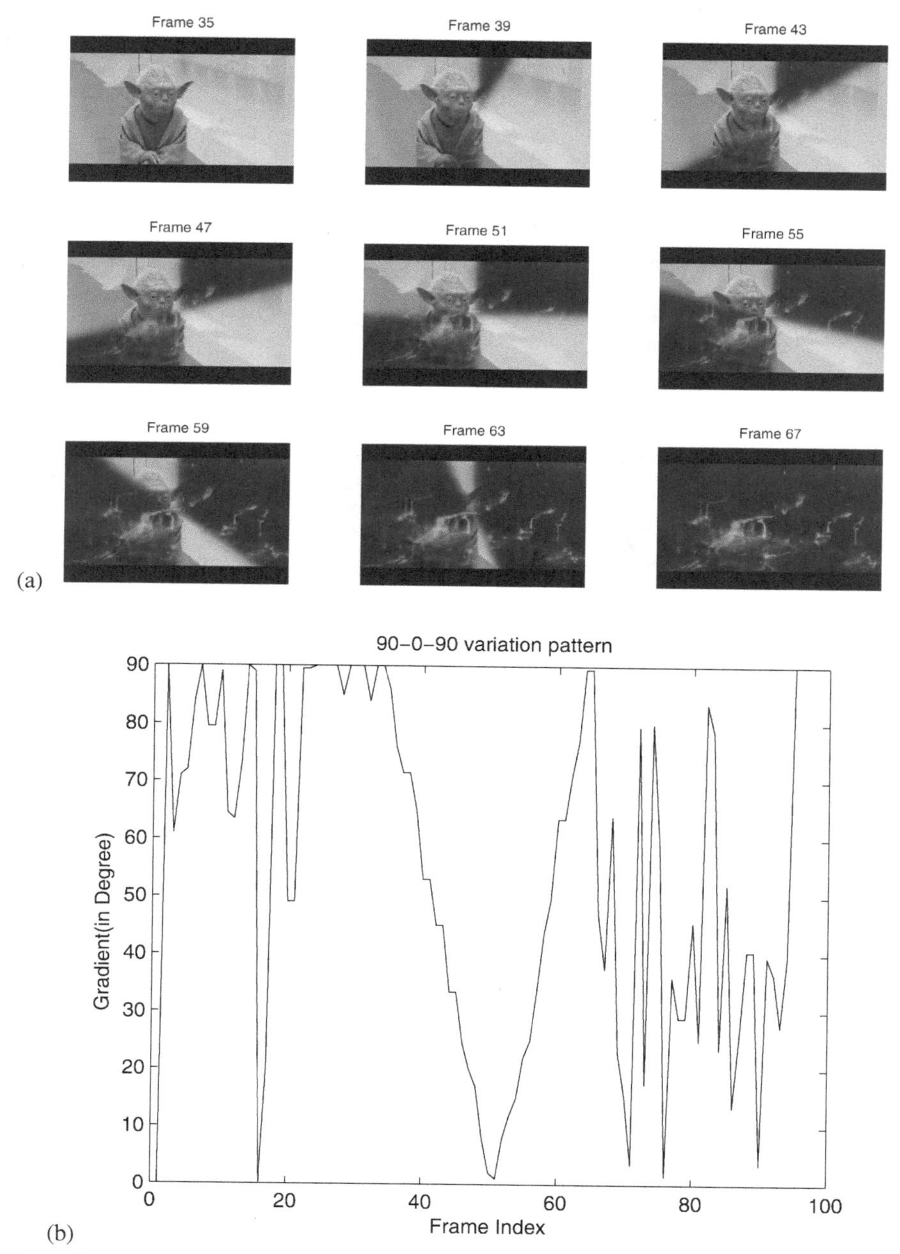

Figure 3.11 (a) Video clip from Star War III movie for special wipe. (b) Gradient pattern satisfied by it

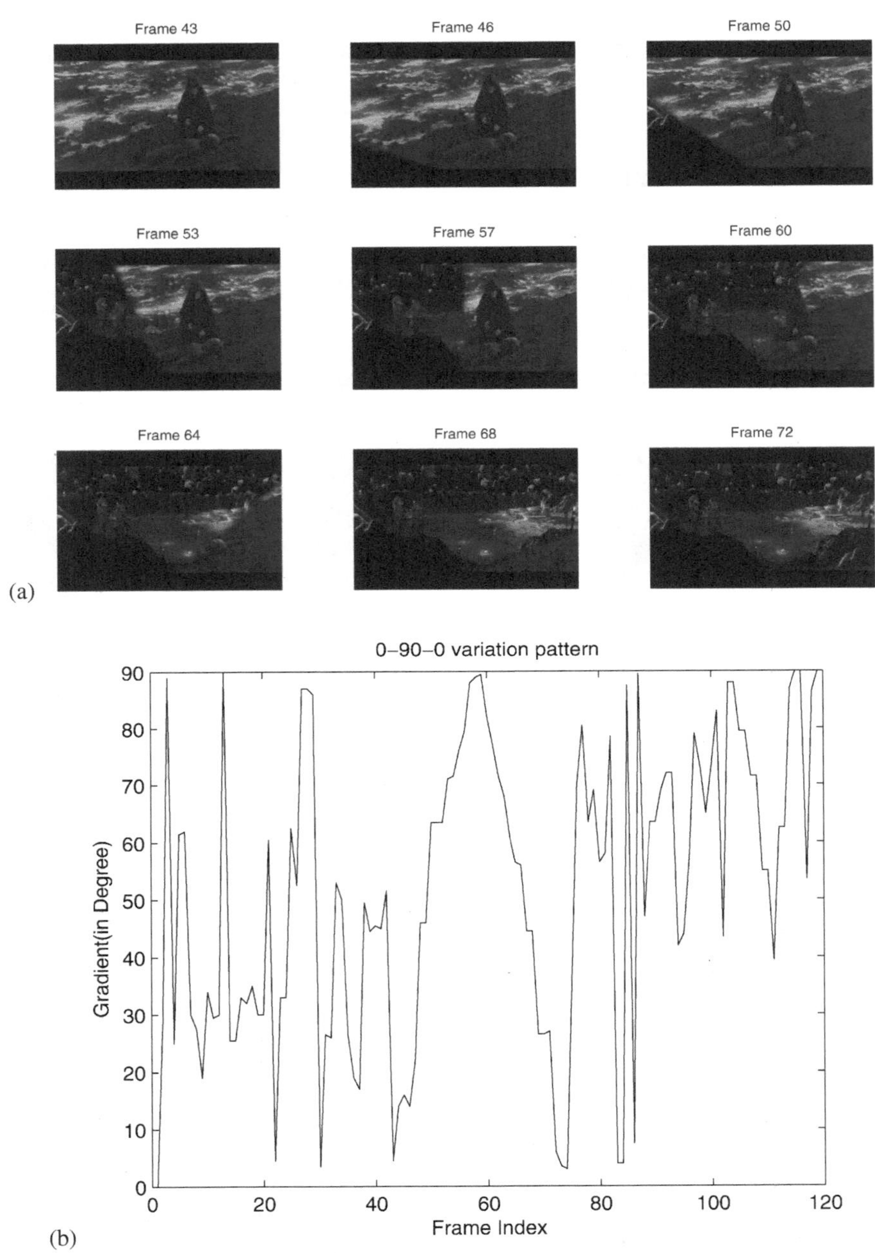

Figure 3.12 (a) Video clip from Star War III movie for special wipe. (b) Gradient pattern satisfies by it

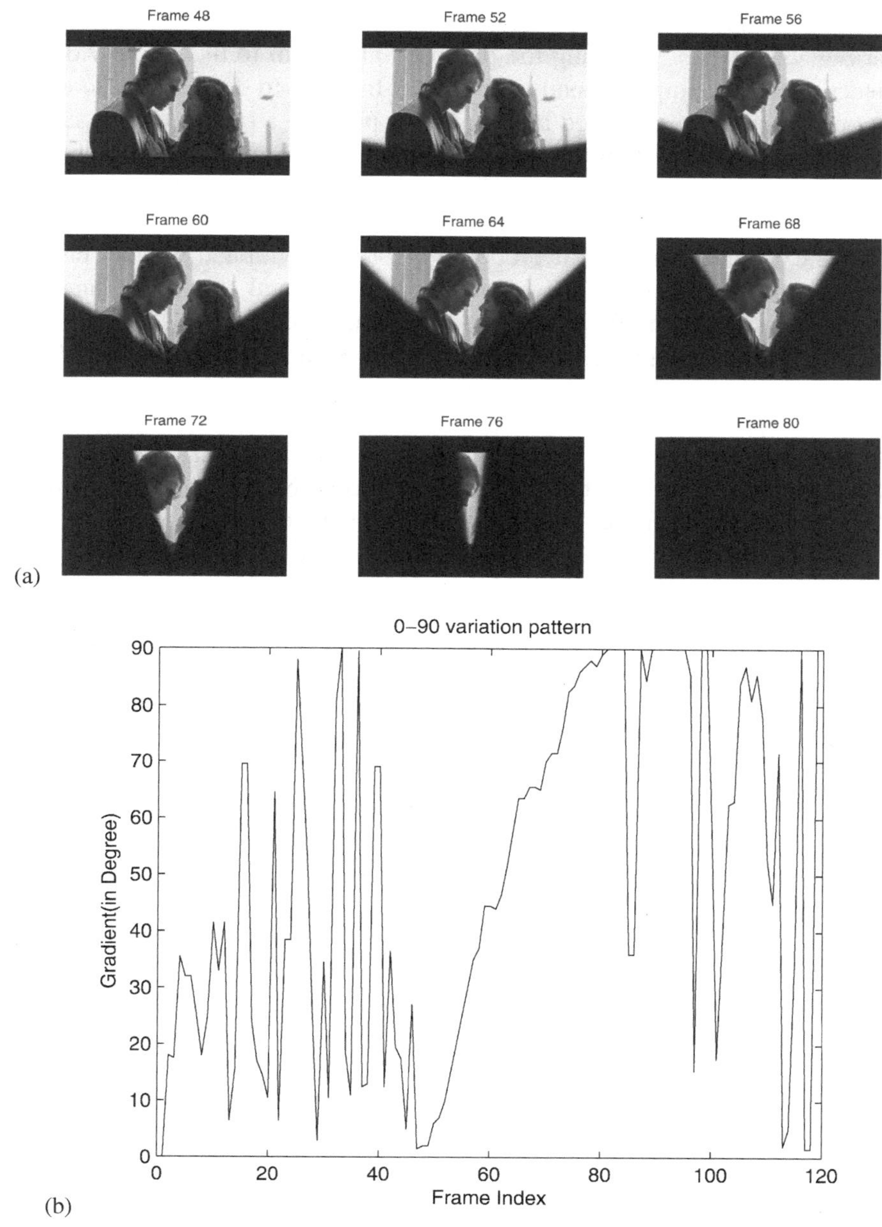

Figure 3.13 (a) Video clip from Star War III movie for special wipe. (b) Gradient pattern satisfied by it

the Hough transform is applied to these moving lines to detect and categorize various wipe types. Applying the proposed algorithm to the whole video to detect wipes is computationally expensive. In order to decrease the computational load of the proposed algorithm, we propose a preprocessing step as the first stage of the algorithm. The potential wipe frames obtained after the preprocessing step are processed through the proposed algorithm to detect wipes.

We extensively tested our proposed algorithm, with and without preprocessing, on several genres of video data. Experimental results are carefully evaluated using the performance metrics Recall, Precision, F1 measure, and Detection Rate. Our proposed algorithm achieved a relatively better trade off between Recall and Precision as compared to other algorithms. We tested the proposed approach on 31 different wipe effects including 18 complex and special wipes, and obtain Detection Rate of 94.12, 100, 100, 100, and 96.67 for Star War III, Star War I, Jodhaa Akbar, Bhooth Nath and Tomcat movies, respectively. The proposed method successfully avoids false positives caused due to object and camera motion, by differentiating wipe transitions and motion using various gradient patterns.

4

Shot Boundary Detection in the Presence of Flashlights

4.1 Introduction and Related Work

Elimination of disturbances caused by flashlight is one of the challenges to the current shot boundary detection. The goal of our work is to take care of flashlights, which is one of the major sources of false positives in shot boundary detection. To address this issue we have proposed an algorithm for shot boundary detection in the presence of flashlight in this chapter. Figure 4.1 shows an example of abrupt transition in the presence of flashlights from the movie Sleepy Hollow.

Related work for shot boundary detection in the presence of flashlights is given below.

Methods for detecting or differentiating flashlight from abrupt transition can be broadly classified as histogram based methods, edge detection based methods and local window based methods. Weixin et al. [24] have used color ratio histogram as illumination invariant metric and local adaptive thresholds are used to find shot change. In [1] histogram difference is used as a metric to differentiate between abrupt transition and gradual transition, and then average intensity difference is used to detect flashlights. This method failed when a flashlight is for a long duration. In [25], Guimaraes et al. showed that flashlight effect forms long white vertical stripes in the visual rhythm by histogram image and these lines are extracted through a white top hat by reconstruction process. White stripes are not clear if flashlight is weak with heavy shadows, and this results in false detection. Yuliang and De [59] used twin comparison algorithm to find potential shot change. By using edge direction histogram and inter frame similarities, light change and shot boundary can be detected. Accumulated histogram difference and energy variation is used to identify flashlights in [60]. This method failed to detect flashlight if starting or ending frames of the flashlights are located at shot boundaries and

Figure 4.1 Video clip having shot boundary in the presence of flashlight

flashlight effect is not due to global luminance change. Such cases can be found in the movie Sleepy Hollow. In [2] frame difference from consecutive frames was detected using subdivided local color histogram comparison. This difference is compressed by using logarithmic transform for efficient detection of flashlights.

Li and Lu [4] proposed a method based on difference in area of background in edge image to differentiate between shot and flashlights. This method assumes that illumination variation is not frequent. In [26] edge direction, edge position, and edge matching is used to discriminate abrupt scene change from flashlight scenes. This method can be computationally intensive and segmentation failed when objects are not clear during flashlights, which is a usual case in horror and thriller movies. Model for flashlight detection is formed by length, strength, brightness, velocity, and impact of the flashlight [58]. Five thresholds are used depending on the above parameters.

Limitations of the existing methods are summarized below:

- Edge detection based methods fail when objects are not clear due to dark background and is computationally expensive.

- Histograms based metrics are sensitive to strong lighting changes but are invariant to small object motion.
- Unable to detect weak flashlight.
- Unable to detect shot boundary when flashlight is present at shot boundary.
- Produces false positives when flashlight is present for long duration which includes weak as well as strong flashlight.
- Produces false positives during switch on/off of a flashlight.
- Failed to detect shot boundaries when objects are not clear during flashlight.

Thus existing flashlight detection methods have limitations in terms of duration of flashlight, nature of flashlight, computation time and threshold. In order to overcome the limitations of the above mentioned algorithms, we developed a technique that suppresses the effect due to flashlight, instead of detecting frames due to flashlight by using a mixture of logarithmic transforms and discrete cosine transforms. Then a discrete wavelet transform based metric, and local or automatic threshold is used to detect shot boundaries. We used color movies as a test data for shot boundary detection. Each color frame is converted to a grayscale in order to reduce the computational load. In [90] it is concluded that color coded global metrics do not provide any significant advantage over the grayscale. Also working on gray scale frames is computationally more efficient than color frames. Majority of changes due to lighting can be captured by luminance variation [8, 25, 58], and therefore we used grayscale frames in our algorithm.

The rest of the chapter is organized as follows. Section 4.2 gives a detailed description of the proposed method to avoid false positives due to flashlight. The effectiveness of the method is validated by experiments on various movies in Section 4.3. Performance comparison with other flashlight detection method is discussed in Section 4.4. Proposed algorithm as a preprocessing stage with other algorithm to improve their results is discussed in Section 4.5. Results under different flashlight conditions using proposed algorithm are shown in Section 4.6.

4.2 Proposed Method to Avoid False Positives Due to Flashlight

In the present chapter, we propose an algorithm which works in three stages as reported in [91]. In the first stage, the illumination effect due to flashlight is suppressed using Logarithmic transform followed by Discrete cosine transform. Second stage involves finding potential shot boundaries using discrete wavelet transform based metric. Finally, in the last stage local or adaptive threshold is used to declare shot boundaries.

4.2.1 Suppression of Illumination Due to Flashlight Using Logarithmic Transform and Discrete Cosine Transform

Illumination invariant processing can be obtained by logarithmically transforming the image and then using a high pass filter, the DC value and the low frequency components are removed while retaining various high frequency features [92]. Logarithmic transform is also used in image enhancement to expand the intensity values of dark pixels. Various methods for illumination invariant image processing are given in [93]. In our proposed algorithm suppression of illumination due to flashlight is obtained by transformation of each image frame into the logarithmic domain followed by discrete cosine transform (DCT) and subsequent removal of the DC component.

Next we explain how illumination due to flashlight is suppressed in our proposed algorithm as reported in [94].

The basic concept for illumination invariance follows from the image formation model in which image intensity of a pixel at a location $[x, y]$ in an image $f[x, y]$, is assumed to be the product of reflectance component $r[x, y]$ and illumination component $l[x, y]$. Consider $f_i[x, y]$ as the current frame with reflectance $r_i[x, y]$ and illumination as $l_i[x, y]$, Then based on the above discussion image formation model is given by

$$f_i[x, y] = l_i[x, y] r_i[x, y] \tag{4.1}$$

Taking the logarithmic transform of both sides in the above equation we get

$$\log(f_i[x, y]) = \log(l_i[x, y]) + \log(r_i[x, y]) \tag{4.2}$$

From the above equation it is clear that by the use of logarithmic transform the multiplicative effect of illumination becomes additive.

Equation (4.2) can be rewritten as

$$\log(r_i[x, y]) = \log(f_i[x, y])) - \log(l_i[x, y]) \tag{4.3}$$

Let $f_{i+1}[x, y]$ be the next consecutive frame with $l_{i+1}[x, y]$ as the illumination component due to flashlight. Normally the reflectance component, which corresponds to the stable features of an image, remains invariant under illumination due to flashlight in the same scene, i.e., $r_{i+1}[x, y] = r_i[x, y]$, and therefore

$$\log(f_{i+1}[x, y]) = \log(l_{i+1}[x, y]) + \log(r_i[x, y]) \quad (4.4)$$

Using Eq. (4.3) we can rewrite Eq. (4.4) as follows:

$$\log(f_{i+1}[x, y]) = \log(f_i[x, y])) + \log(l_{i+1}[x, y]) - \log(l_i[x, y]) \quad (4.5)$$

Defining the term

$$C[x, y] = \log(l_{i+1}[x, y]) - \log(l_i[x, y]) \quad (4.6)$$

as the 'flashlight disturbance', Eq. (4.5) can be rewritten as

$$\log(f_{i+1}[x, y]) = \log(f_i[x, y]) + C[x, y] \quad (4.7)$$

From the above equation it is clear that, in order to eliminate the effect of flashlight, it is essential to suppress the term $C[x, y]$ so that in the same scene, the next frame term $\log(f_{i+1}[x, y])$ becomes almost equal to the current frame term $\log(f_i[x, y])$.

We next discuss the procedure to suppress $C[x, y]$ with the use of DCT. We used DCT-II for transformation since it can be implemented using fast algorithm to reduce computational complexity.

The desired number of frames from the video is captured. Each frame is converted from RGB to YUV color space, and only luminance (Y) component is used to obtain gray scale frame for further analysis. This grayscale frame is transformed into logarithmic domain by taking its logarithmic transform to obtain *log-image*. Then 2D-DCT is used to transform this logarithmic image, i.e., *log-image* into frequency domain to obtain the *DCT-log-image*.

The general equation for 2D ($M \times N$ image) DCT is defined by the following equation:

$$C(u, v) = \alpha(u)\alpha(v) \sum_{x=0}^{M-1} \sum_{y=0}^{N-1} f(x, y) \cos\left[\frac{\pi(2x+1)u}{2M}\right] \cos\left[\frac{\pi(2y+1)v}{2N}\right] \quad (4.8)$$

and the inverse 2D-DCT is defined as

$$f(x, y) = \sum_{u=0}^{M-1} \sum_{v=0}^{N-1} \alpha(u)\alpha(v)C(u, v) \cos\left[\frac{\pi(2x+1)u}{2M}\right] \cos\left[\frac{\pi(2y+1)v}{2N}\right] \quad (4.9)$$

where

$$\alpha(u) = \begin{cases} \frac{1}{\sqrt{M}} & \text{for } u = 0 \\ \sqrt{2/M} & \text{for } u = 1, 2, \ldots, M-1 \end{cases}$$

$$\alpha(v) = \begin{cases} \frac{1}{\sqrt{N}} & \text{for } v = 0 \\ \sqrt{2/N} & \text{for } v = 1, 2, \ldots, N-1 \end{cases}$$

Any pixel in an image is considered to be the product of reflectance and illumination at that point. Illumination variation is usually smooth in spatial domain and, therefore, this corresponds to low frequency components in the transform domain, whereas reflectance variation is usually rapid, and this results in high frequency components in the transform domain. The first DCT coefficient, also known as DC coefficient, represents the average intensity of the image and is most affected by variation in illumination component in the image.

Due to logarithmic transform the multiplicative effect of illumination in the image becomes additive, and therefore, when the logarithmic image is transformed into frequency domain using DCT, the illumination component is mostly reflected in the DC coefficient and the low frequency coefficients of the *DCT-log-image*.

The flashlight disturbance, i.e., $C[x, y]$ is a difference of illumination due to flashlight and normal illumination. Normally both these illuminations have nearly uniform spatial distribution, except that the illumination due to flashlight will have higher intensity compared to the normal illumination. Therefore the DCT of $C[x, y]$ will contribute significantly to the DC value in the *DCT-log-image*. Consequently, the removal of the DC value in the *DCT-log-image* suppress flashlight disturbance, i.e., the term $C[x, y]$ in Eq. (4.7). However, if illumination due to flashlight is nonuniform or varying rapidly in spatial domain then the DCT of $C[x, y]$ will contribute to the DC and other frequency component in the *DCT-log-image*. In this case the removal of the DC coefficient will still suppress the illumination due to flashlight but will be less effective.

4.2.2 Wavelet Transform as a Shot Detection Matrix

After suppressing illumination each frame is now processed with a discrete wavelet transform. The multiresolution nature of wavelet analysis provide a compact representation of various types of signal localized in space, time or frequency domain.

Since we are using separable wavelets, a 2D wavelet transform is computed by iterating above filter row wise and then column wise. The filters used in this decomposition are the first order Daubechies low pass and high pass, quadrature mirror filters [95]. Such a process is recursively performed on the low band, leading to a hierarchical pyramid type decomposition. When each frame data in a video sequence is decomposed by a 2D wavelet transform up to the scale of three, its hierarchical properties provide a new scalable, extensible representation of video data. This decomposition reduces the computational load of the proposed algorithm. Our extensive investigation results shows that decomposition to a higher level than three leads to substantial loss of information.

Let a frame data of original resolution be $X(i, j)$ and its wavelet decomposition up to the scale of 2^{-J} be $X(i, j) \rightarrow \{LL3, (LH3, HL3, HH3)\}$, where $J = 3$ and LL3 is a coarse version of the original image. For all the frames the coarse version is represented by $\{LL3(i, j, 1), LL3(i, j, 2), \ldots, LL3(i, j, n)\}$, where n is the number of frames used for the analysis. Then the wavelet difference between consecutive frames by considering this low resolution low scale subimages is given by

$$WDBF(k) = \sum_{i=1}^{P}\sum_{j=1}^{Q}[LL3(i, j, k) - LL3(i, j, k+1)]^2 \qquad (4.10)$$

for $1 \leq k \leq n-1$, where $P = M/2^J$, $Q = N/2^J$ and n is the number of frames in the video.

4.2.3 Threshold Selection

Threshold selection is a key issue in video segmentation. Global thresholds are not sufficient as video property could change dramatically when content changes and it is often impossible to find universal threshold. Because of suppression of flashlight, applying threshold becomes easy in our algorithm.

Local Threshold After experimenting on the large data set, we determined the local threshold (L_T) as

$$L_T = \alpha \times \frac{1}{n-1}\sum_{l=1}^{n-1} WDBF(l) + \beta \times \sqrt{\frac{1}{n-1}\sum_{l=1}^{n-1}\left[WDBF(l) - \frac{1}{n-1}\sum_{l=1}^{n-1} WDBF(l)\right]^2} \quad (4.11)$$

where α and β are tuning parameters and can be tuned for better trade-off between Recall and Precision as discussed in Section 4.3.2.

Adaptive Threshold The local threshold depends on the tuning parameter and has to be tuned for trade-off between Recall and Precision. Hence we defined an adaptive threshold (A_T) which is independent of any tuning parameters and is defined in Eq. (4.12). AI is the average of WDBF coefficients in the left sliding window (which is 7 frames preceding the WDBF value of the current frame), WDBF coefficients in the right sliding window (which is 7 frames after the WDBF value of the current frame) and WDBF value of the current frame. Therefore AI is the addition of mean and standard deviation of WDBF for 15 frames including the current frame for various intervals described in Eq. (4.12). For the first 7 frames the left sliding window does not exist, and hence, only the average WDBF value of the right sliding window is considered, whereas for the last 7 frames the right sliding window does not exist, hence only the average WDBF value of the left sliding window is considered:

$$A_T(k) = \frac{1}{n-1}\sum_{l=1}^{n-1} WDBF(l) + \sqrt{\frac{1}{n-1}\sum_{l=1}^{n-1}[WDBF(l) - \frac{1}{n-1}\sum_{l=1}^{n-1} WDBF(l)]^2 + AI(k)} \quad (4.12)$$

where

$$AI(k) = \begin{cases} \frac{1}{15}\sum_{l=1}^{15} WDBF(l) \\ \quad + \sqrt{\frac{1}{15}\sum_{l=1}^{15}[WDBF(l) - \frac{1}{15}\sum_{l=1}^{15} WDBF(l)]^2} \\ \quad \text{for } 1 \leq k \leq 7 \\ \frac{1}{15}\sum_{l=k-7}^{k+7} WDBF(l) \\ \quad + \sqrt{\frac{1}{15}\sum_{l=k-7}^{k+7}[WDBF(l) - \frac{1}{15}\sum_{l=k-7}^{k+7} WDBF(l)]^2} \\ \quad \text{for } 8 \leq k \leq n-9 \\ \frac{1}{15}\sum_{l=n-15}^{n-1} WDBF(l) \\ \quad + \sqrt{\frac{1}{15}\sum_{l=n-15}^{n-1}[WDBF(l) - \frac{1}{15}\sum_{l=n-15}^{n-1} WDBF(l)]^2} \\ \quad \text{for } n-8 \leq k \leq n-1 \end{cases}$$

4.2.4 Shot Boundary Detection Algorithm

1. The desired number of frames from the video is captured. Each frame is converted from RGB to YUV color space, and only luminance component is used for further analysis.
2. Let $s(x, y, t)$ represent the grayscale value at location (x, y) obtained by converting each color frame to grayscale for $1 \leq t \leq n$.
3. Each frame is transformed into a logarithmic domain by $s_L(x, y, t) = \log(s(x, y, t))$ for $1 \leq t \leq n$.
4. Each frame $s_L(x, y)$ is transformed into a frequency domain by

$$C(u, v) = \alpha(u)\alpha(v)\sum_{x=0}^{M-1}\sum_{y=0}^{N-1} \times s_L(x, y)\cos\left[\frac{\pi(2x+1)u}{2M}\right]\cos\left[\frac{\pi(2y+1)v}{2N}\right]$$

5. Set the first DCT coefficient of $C(u, v)$ to zero in all frames.

6. Find inverse 2D-DCT of all frames after removing the DCT coefficient by

$$X(i, j) = \sum_{i=0}^{M-1} \sum_{j=0}^{N-1} \times \alpha(u)\alpha(v)C(u, v) \cos\left[\frac{\pi(2i+1)u}{2M}\right] \cos\left[\frac{\pi(2j+1)v}{2N}\right]$$

7. We decompose each frame $X(i, j)$ up to the scale of $J = 3$, and its coarse version at this scale is represented by $\{LL3(i, j, 1), LL3(i, j, 2), \ldots, LL3(i, j, n)\}$, where n is the number of frames used for the analysis. Then wavelet difference between consecutive frames by considering this low resolution low scale subimages is calculated by

$$WDBF(k) = \sum_{i=1}^{P} \sum_{j=1}^{Q} [LL3(i, j, k) - LL3(i, j, k+1)]^2$$

for $1 \leq k \leq n - 1$, where $P = M/2^J$ and $Q = N/2^J$.
8. Calculate the local threshold L_T and adaptive threshold $A_T(k)$ by Eqs. (4.11) and (4.12) respectively.
9. For $1 \leq k \leq n - 1$

{if [$WDBF(k)$ > Threshold
declare camera break}

Threshold L_T is used if results are to be obtained using local threshold and threshold $A_T(k)$ is used if results are to be obtained using an adaptive threshold.

4.3 Experimental Results and Discussion

4.3.1 Test Video Sequence and an Evaluation Criterion

To evaluate the effectiveness of the proposed algorithm, experiments are conducted on horror and thriller movies such as 'Sleepy Hollow' (SH), 'Independence Day' (ID), and 'Deep Rising' (DR). All these movies have the following characteristics in common: a dark background, a long flashlight duration, and a combination of weak and strong flashlights. To test the robustness of the algorithm, we consider frames from the video where flashlights

Table 4.1 Tuning parameters and its effect on Recall and Precision

TP		SH			ID			DR		
α	β	R	P	F1	R	P	F1	R	P	F1
2	2	97.67	89.36	93.33	100	85	91.89	93.33	58.33	71.79
2	2.5	97.67	93.33	95.45	100	89.47	94.44	93.33	70	79.99
2	3	88.37	92.68	90.47	100	94.44	97.14	93.33	87.5	90.32
2	**3.5**	79.07	91.89	84.99	**100**	**94.44**	**97.14**	93.33	87.5	90.32
2	**4**	67.44	93.55	78.37	82.35	93.33	87.49	**93.33**	**93.33**	**93.33**
2.5	2	97.67	89.36	93.33	100	85	91.89	93.33	58.33	71.79
3	2	97.67	93.33	95.45	100	89.47	94.44	93.33	66.67	77.77
3.5	**2**	**97.67**	**97.67**	**97.67**	100	89.47	94.44	93.33	70	79.99
4	2	97.67	93.33	95.45	100	89.47	94.44	93.33	70	79.99
1	1	97.67	44.21	60.86	100	48.57	65.38	100	30	46.15
4	4	60.47	92.86	73.24	82.35	93.33	87.49	86.67	100	92.85

are more than camera breaks (shot boundary), with a combination of strong and weak flashlights. For example in the movie Sleepy Hollow, the portion of the video that we considered for simulation has 43 frames with camera breaks and 401 frames with flashlight in that duration, followed by 17 frames with camera break and 482 frames with flashlight in Independence Day. In the movie Deep Rising, the portion of the video considered has 15 frames with camera break and 232 frames with flashlight. All these selected videos have been manually observed frame by frame to find frames with camera breaks and frames with flashlights.

We used Recall, Precision and F1 measure as discussed in Section 2.5 as an evaluation metric to compare the shot boundary detection algorithm.

4.3.2 Selection of Tuning Parameters and Its Effect on Recall and Precision

We extensively varied the values of the tuning parameter in Eq. (4.11) (TP) and tested various movie videos for Recall and Precision. It has been empirically observed that for the range $2 \leq \alpha \leq 4$ and $2 \leq \beta \leq 4$, trade-off between Recall and Precision is better. If lower values of tuning parameters are used, then Precision drops considerably, whereas for higher values of these parameters Recall degrades. Details about the variation of α and β parameter and its effect on Recall and Precision are shown in Table 4.1.

For best Recall and Precision for different movie videos, the selected values of α and β parameters are shown in Table 4.2. The use of tuning parameter is sometimes necessary, because it plays an important role in the segmentation

Table 4.2 Selected values of α and β for different movie videos

Video	Sleepy Hollow	Independence Day	Deep Rising
α	3.5	2	2
β	2	3.5	4

Table 4.3 Performance comparison of the proposed algorithm for different movies with [1–4]. D: desired detection, C: correct detection, M: missed detection, FP: false positive, R: Recall, P: Precision, NFRWL: number of frames with flashlight in the test video

Algorithm	Video	D	C	M	FP	R	P	F1	NFRWFL
Proposed	SH	43	42	01	01	97.67	97.67	97.67	401
	ID	17	17	00	01	100	94.44	97.14	482
	DR	15	14	01	01	93.33	93.33	93.33	232
Zhang et al. [1]	SH	43	24	19	38	55.88	38.70	45.72	401
	ID	17	08	09	46	47.05	14.81	22.52	482
	DR	15	02	13	26	13.33	7.14	9.29	232
Cheol et al. [2]	SH	43	31	12	34	72.09	47.69	57.40	401
	ID	17	09	08	13	52.94	40.90	46.14	482
	DR	15	05	10	01	33.33	83.33	47.61	232
Zhang et al. [3]	SH	43	23	20	44	53.48	34.32	41.80	401
	ID	17	04	13	32	23.52	11.11	15.09	482
	DR	15	02	13	27	13.33	6.89	9.08	232
Li et al. [4]	SH	43	23	20	10	53.48	69.69	60.51	401
	ID	17	13	04	07	76.47	65	70.27	482
	DR	15	13	02	07	86.66	66.66	75.35	232

process where the user can suitably vary it according to the nature of data and the type of application. Also, the tuning of this parameter allows us to find a compromise between over-segmentation and under-segmentation.

4.4 Performance Comparison with the Different Flashlight Detection Methods

4.4.1 Performance Comparison

We compare the performance comparison of the proposed algorithm with algorithm by Zhang et al. [1], Cheol et al. [2], Zhang et al. [3], and Li et al. [4] for the same video sequences. Detection results of these algorithms for each test video sequences are shown in Table 4.3.

The algorithm by Zhang et al. [1] proposes a flash model and cut model to differentiate between abrupt changes and flashlight. This algorithm assumes that flashlight event do not last longer than 7 frames, and it returns to original state after flashlight. In the test video sequences that we considered, there are many instances where flashlight exceed more than 7 frames, and sequence does not return to the original state after flashlight. These are the main causes for failure of this method. The method proposed by Cheol et al. [2] is reasonably robust against flashlight. In [2], the unappropriate threshold selection is the main cause for low Recall and Precision. The algorithm proposed by Zhang et al. [3], which is popularly used for shot detection, provides poor performance, because this algorithm is not robust for illumination change, and in our test video sequences we have considerable number of frames with flashlight. The algorithm proposed by Li and Lu [4] performs reasonably well as compared to other tested algorithms [1–3], since it is invariant to flashlight. This algorithm assumes that the number of edges disappearing or appearing is very limited during the occurrence of the flashlight, and it counts the number of pixels contributing for computation of ABEM (Area of Background in Edge iMage), which is sometimes not the case in real videos. Threshold selection is another limitation of this algorithm.

The above discussed algorithms either use known metrics or modified metrics for shot boundary detection. In our proposed method, instead of using these known metrics, we use mixture of transforms to suppress illumination caused by flashlight. The efficacy of the proposed method can be clearly observed in Table 4.3; the proposed method outperforms other methods in terms of better Recall and Precision. The proposed algorithm is also robust in the sense that our test data contain frames with flashlight 10–30 times more than frames with abrupt transition and contain a combination of weak and strong flashlights. Other methods used for comparison are unable to avoid false positives caused by flashlight on the same data, whereas our method provides better trade-off between Recall and Precision. However, the performance of the proposed method is sensitive to large camera/object motion and results in false positives.

The main goal of the chapter is to suppress the disturbances caused by flashlight which is one of the major challenges in shot boundary detection and we achieved it reasonably well. Our test data contain mostly flashlight with camera breaks and reasonably small object/camera motion (observed in the movie Sleepy Hollow) and few frames with large camera/object motion (observed in the movie Independence). The Illumination Suppression algorithm using logarithmic transform and DCT has resulted in suppression of flashlights, whereas wavelet transform based metrics in combination with adaptive threshold have been successful in avoiding the effect of small camera and object motion.

Table 4.4 Performance comparison of the proposed algorithm using the local threshold and an adaptive threshold

Proposed algorithm	Video	D	C	M	FP	R	P	F1
Results using local threshold	SH	43	42	01	01	97.67	97.67	97.67
	ID	17	17	00	01	100	94.44	97.14
	DR	15	14	01	01	93.33	93.33	93.33
Results using adaptive threshold	SH	43	42	01	02	97.67	95.45	96.54
	ID	17	17	00	02	100	89.47	94.44
	DR	15	15	00	01	100	93.33	96.54

Table 4.5 Speed comparison of the different algorithm

Algo.	Zhang et al. [3]	Zhang et al. [1]	Li and Lu [4]	Proposed	Cheol et al. [2]
Speed	8.124	10.996	18.1248	125	790

4.4.2 Performance Comparison of the Proposed Algorithm Using the Local Threshold and an Adaptive Threshold

We extensively varied the values of tuning parameter α and β in local threshold and tested results on various movie for Recall and Precision. It has been empirically observed that for the range $2 \leq \alpha \leq 4$ and $2 \leq \beta \leq 4$, trade-off between Recall and Precision is better. The values of α and β slightly change for different video sequences as shown in Table 4.2 but remains in the range $2 \leq \alpha(\beta) \leq 4$ for best results.

To make results insensitive to these tuning parameters, we defined an adaptive threshold as shown in Eq. (4.12). Experimental results are carried on the SH, ID and DR movies using these automatic thresholds in the proposed algorithm and results are shown in Table 4.4. It has been found that Recall remain same for SH and ID, whereas it improves for DR movie as one of the shot boundary missed due to the local threshold is detected by an adaptive threshold. Precision goes slightly low in SH and ID movie due to one additional false positive as compared to local threshold, but remains same for DR movie. Overall by using the adaptive threshold, F1 score drops by 1 and 3% in the SH and ID movies, whereas it improves by 3% in the DR movie.

4.4.3 Computational Time for the Different Detection Methods

All the algorithms were implemented using Matlab 7.0. Average computational time was obtained by running the program on Intel P4, CPU 2.40 GHz with 512 MB RAM. Table 4.5 shows computational time for the different detection methods, where speed is in milliseconds per frame.

Table 4.6 Illumination suppression method as a preprocessing step with Li and Lu [4]

Algorithm	Video	D	C	M	FP	R	P	F1	NFRWFL
Li and Lu [4]	SH	43	23	20	10	53.48	69.69	60.51	401
	ID	17	13	04	07	76.47	65	70.27	482
	DR	15	13	02	07	86.66	66.66	75.35	232
Proposed with Li	SH	43	31	12	07	72.09	81.58	76.54	401
	ID	17	15	02	05	88.24	75	81.08	482
	DR	15	14	01	05	93.33	73.68	82.34	232

Although the computation cost of the proposed algorithm is higher compared to other algorithms, the results are better in terms of Recall and Precision. One of the problems with all shot detection algorithms are, if more accurate algorithm is used, more time will be spent on the detection. In our algorithm, we focused on accuracy assuming there is sufficient computational power available. Hence there is a trade-off between accuracy and computational time which has been investigated by Mikolay et al. [96].

4.5 Proposed Suppression Algorithm as a Preprocessing Stage with Other Algorithms to Improve Results

The first stage of our proposed method which is used to suppress flashlight effect using logarithmic transform and discrete wavelet transform can be a preprocessing step for other metrics and may reduce the false positives due to flashlights. We applied steps 1 to 6 from the shot boundary detection algorithm (Section 4.2.4) as a preprocessing step for the algorithm proposed by Li and Lu [4] and the results are shown in Table 4.6. It can be clearly seen that the F1 score increases by 16, 11, and 4% respectively by using the first stage of our proposed method as a preprocessing step for this algorithm. These results show that suppression algorithm if used as a preprocessing step for other detection methods improves the detection performance.

4.6 Results under Different Flashlight Conditions Using the Proposed Algorithm

4.6.1 Shot Boundary Detection during Weak Flashlight

Figure 4.2 shows frames from the movie Sleepy Hollow during weak flashlight condition, where shot boundaries are at frame numbers 21, 50, and 118. Figure 4.3(a) shows wavelet difference between frames without proposed

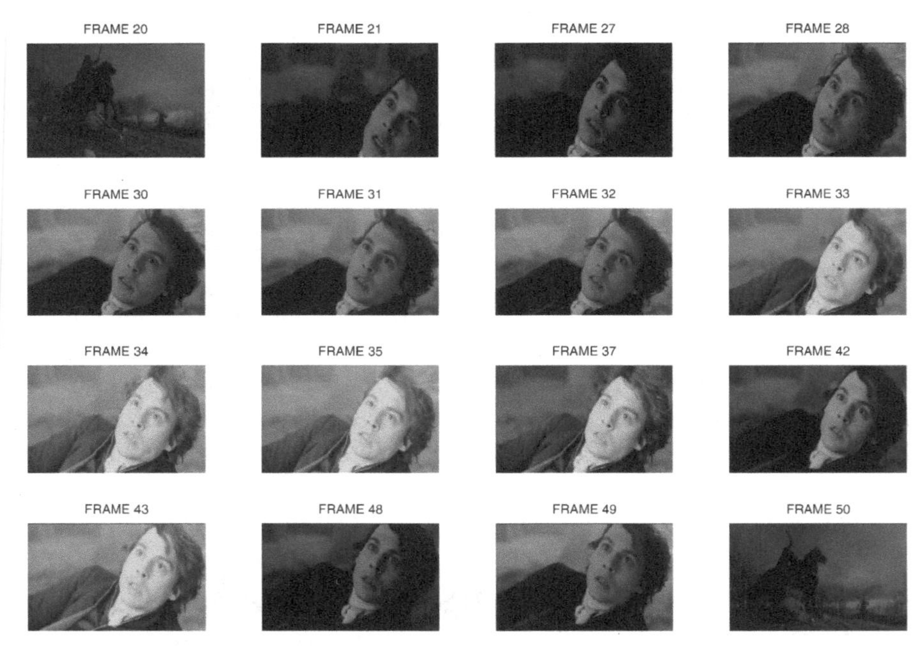

Figure 4.2 Video clip from the Sleepy Hollow movie in weak flashlight condition (true shot boundary at frame numbers 21, 50, and 118)

algorithm. If threshold is applied to this frame difference to detect all shot boundaries mentioned above, then a large number of false positives will be detected due to flashlight effect. Figure 4.3(b) shows the results after applying the proposed algorithm, which clearly indicate that the effect due to flashlight is suppressed, and all shot boundaries are easily detected after applying threshold. The threshold is shown by dotted lines in Figure 4.3(b).

4.6.2 Shot Boundary Detection When Flashlights Are Present in Shot Boundary

Figure 4.4 shows frames from the movie Sleepy Hollow where flashlights are present during shot boundary. Here shot boundaries are present at frame numbers 21 and 55 and also flashlights are present during transition from one scene to another scene i.e. at frame numbers 20 and 55. Figure 4.5(a) shows the wavelet difference between the frames without the proposed algorithm. If threshold is applied to this frame difference to detect all shot boundaries including at the 55th frame, then false positives will be detec-

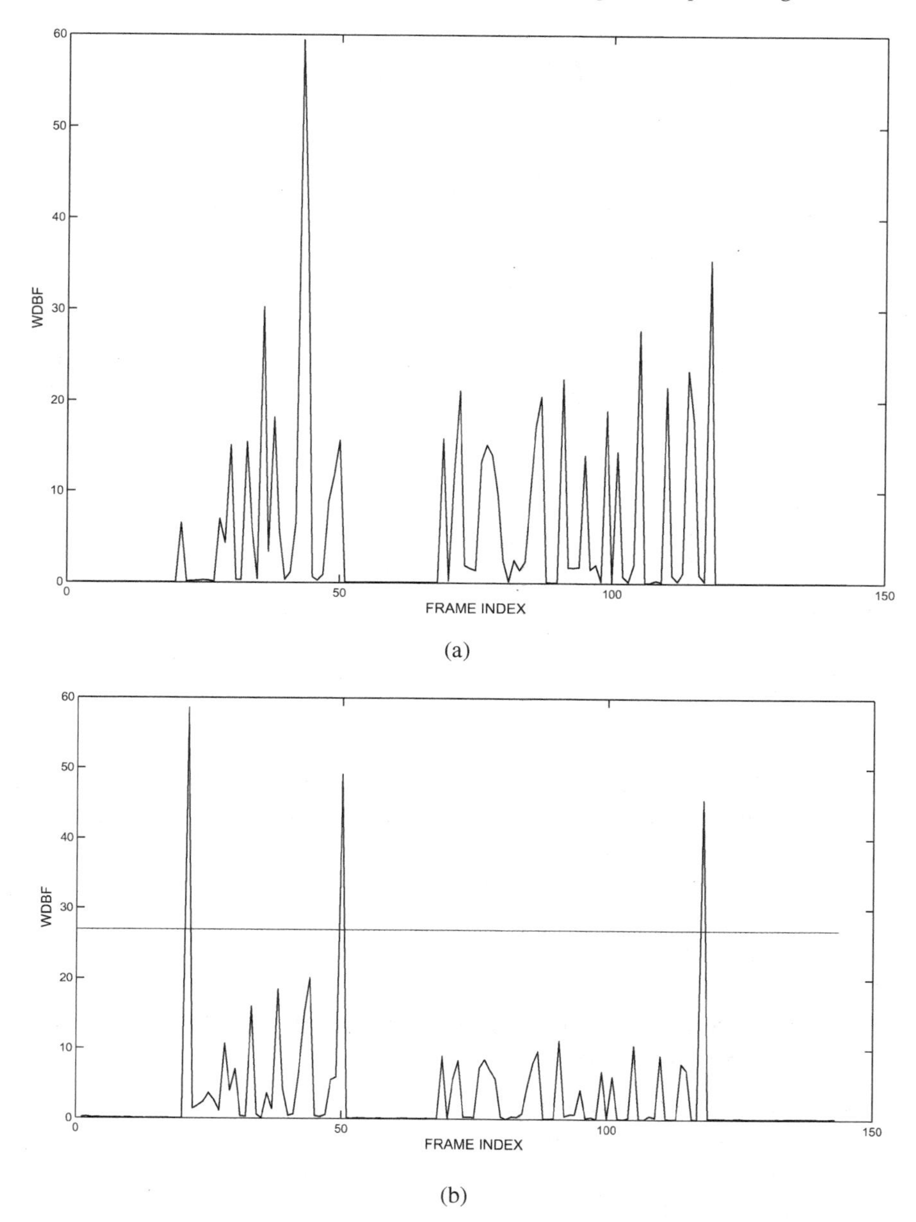

Figure 4.3 (a) Wavelet difference between frames without the proposed algorithm for the video clip (shown in Figure 4.2). (b) Wavelet difference between frames with the proposed algorithm for the video clip (shown in Figure 4.2) (true shot boundaries are at frame numbers 21, 50, and 118)

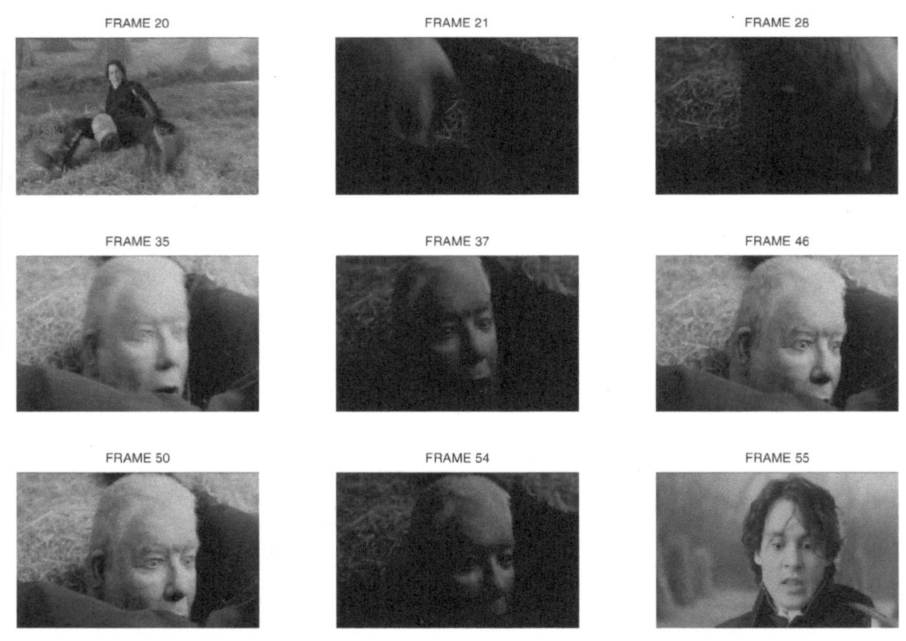

Figure 4.4 Video clip from the Sleepy Hollow movie when flashlight is present in shot boundary (true shot boundary at frame numbers 21 and 55)

ted due to flashlight effect. Figure 4.5(b) shows results after applying the proposed algorithm, which clearly shows that if flashlight is present during shot boundary, still all shot boundaries are detected after applying threshold. This becomes possible thanks to the flashlight suppression algorithm which suppressed the effect due to flashlight and results in avoiding false positives. The threshold is shown by dotted lines in Figure 4.5(b).

4.6.3 Shot Boundary Detection When Flashlights Are Present for Long Duration

Figure 4.6 shows frames from the movie Sleepy Hollow during long flashlight condition, where shot boundary is at frame number 28. Figure 4.7(a) shows wavelet difference between frames without the proposed algorithm. Here peaks from frame number 40–52 is due to long flashlight which includes weak as well as strong flashlights, and peak at frame number 67 is due to strong flashlight. If threshold is applied to this frame difference to detect shot boundary at frame number 28, then false positives will be detected at frame

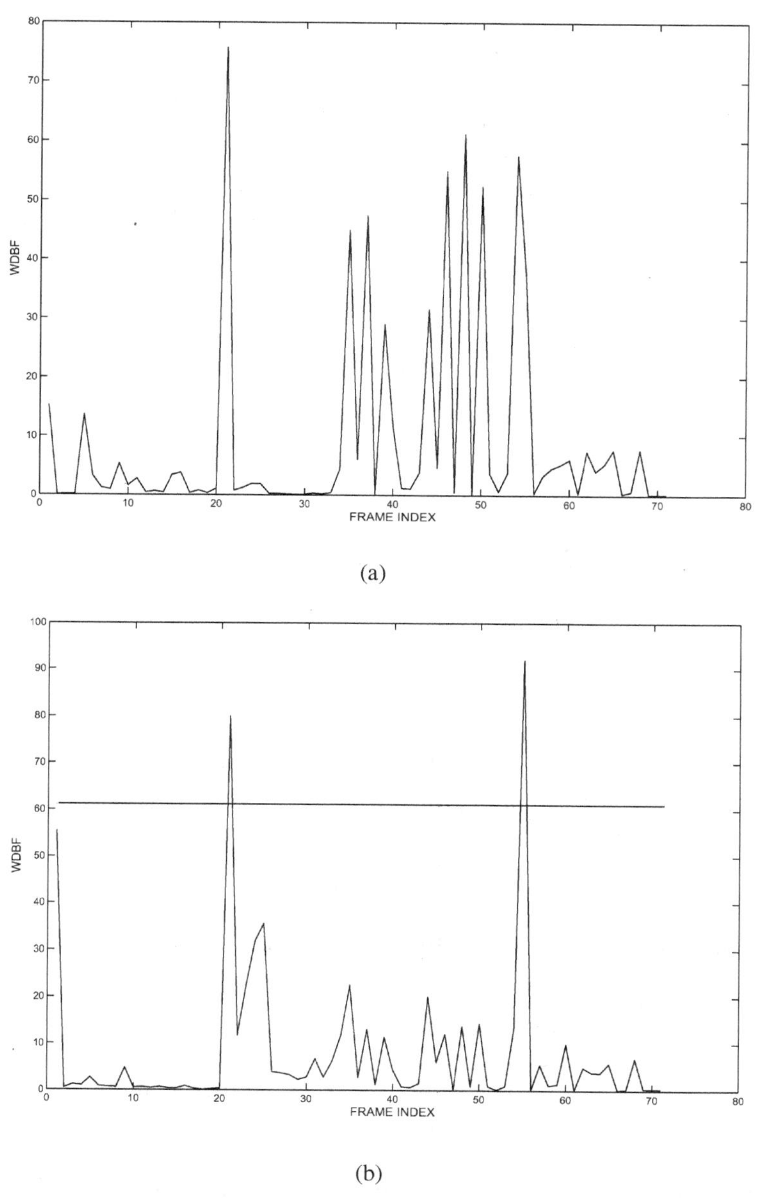

Figure 4.5 (a) Wavelet difference between frames without the proposed algorithm for the video clip (shown in Figure 4.4). (b) Wavelet difference between frames with the proposed algorithm for the video clip (shown in Figure 4.4) (true shot boundary at frame numbers 21 and 55)

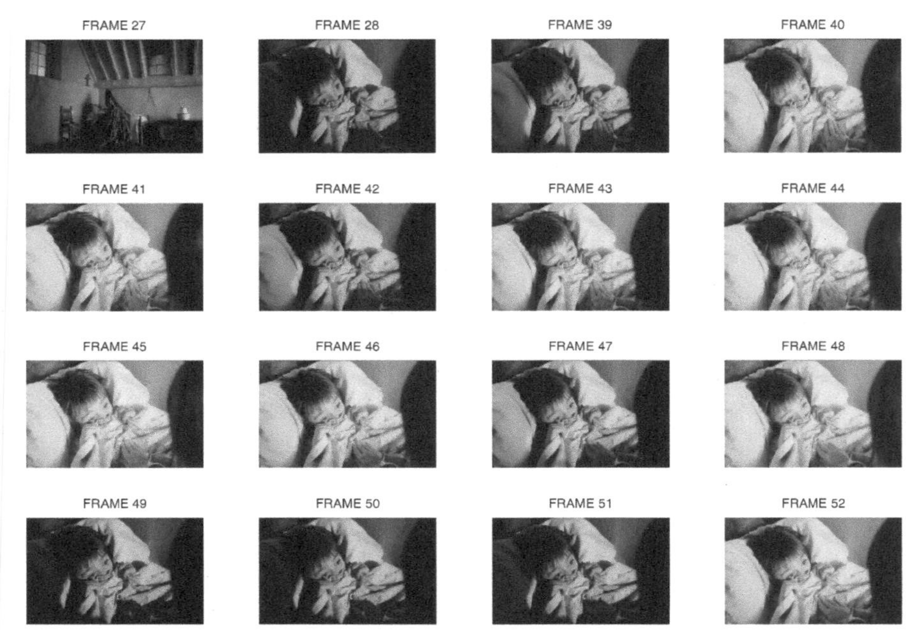

Figure 4.6 Video clip from the Sleepy Hollow movie when flashlights are present for long duration (true shot boundary at frame number 28)

numbers 40, 42, 48, 49, 52, and 67 due to flashlight effect. Figure 4.7(b) shows the results after applying the proposed algorithm, where peaks due to flashlight is suppressed, and shot boundary at frame number 28 is detected. The threshold is shown by dotted lines in Figure 4.7(b).

4.6.4 Shot Boundary Detection When Objects Are Not Clear during Flashlights

Figure 4.8 shows frames from the movie Deep Rising where objects are not clear during flashlight. Here shot boundary is at frame number 59. Figure 4.9(a) shows wavelet difference between frames without the proposed algorithm. As can be seen from Figure 4.8, many frames have a dark background before and/or after flashlight; also when flashlight is present, objects are not clear. This difference in background results in a large wavelet difference between frames 2–45. If the threshold is applied to this frame difference to detect shot boundary at frame number 59, then a large number of false positives will be detected. Figure 4.9(b) shows the results after applying

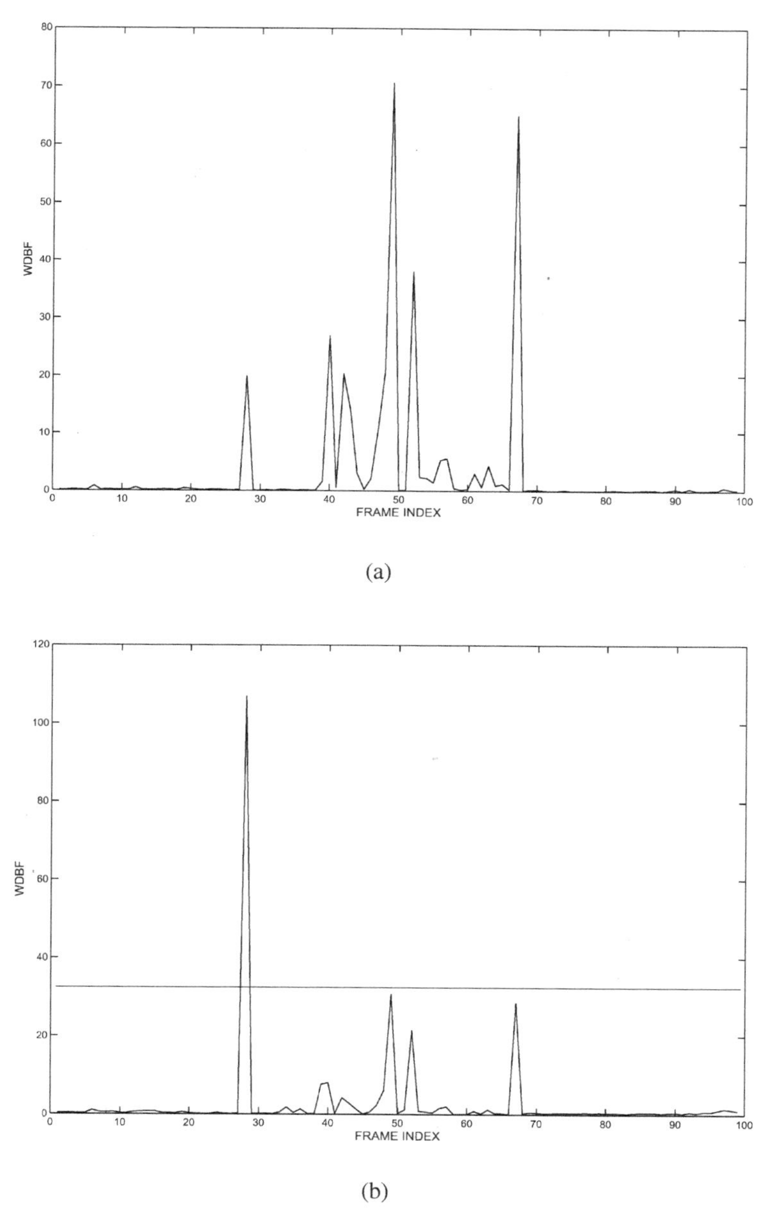

Figure 4.7 (a) Wavelet difference between frames without the proposed algorithm for the video clip (shown in Figure 4.6). (b) Wavelet difference between frames with the proposed algorithm for the video clip (shown in Figure 4.6) (true shot boundary at frame number 28)

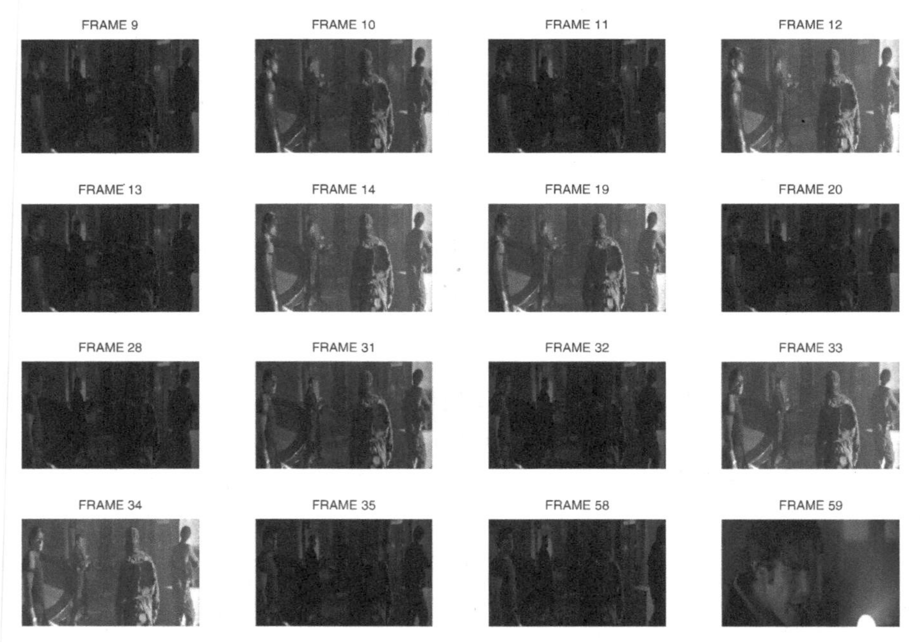

Figure 4.8 Video clip from the Deep Rising movie when objects are not clear during flashlight (true shot boundary at frame number 59)

the proposed algorithm, where peaks due to flashlight are suppressed, and shot boundary is detected at frame number 59 after applying threshold. The threshold is shown by dotted lines in Figure 4.9(b).

4.7 Summary and Conclusions

In this chapter an effective method for shot boundary detection in the presence of flashlights is presented. The proposed algorithm suppresses the effect due to flashlight by using a logarithmic transform followed by a discrete cosine transform. Discrete wavelet transform based metric in combination with local and automatic thresholds is used to find shot boundaries. Comparison of results with other shot boundary detection methods in the presence of flashlight for various movie sequence shows the effectiveness of the proposed algorithm. The proposed method outperforms other compared methods with better trade-off between Recall and Precision.

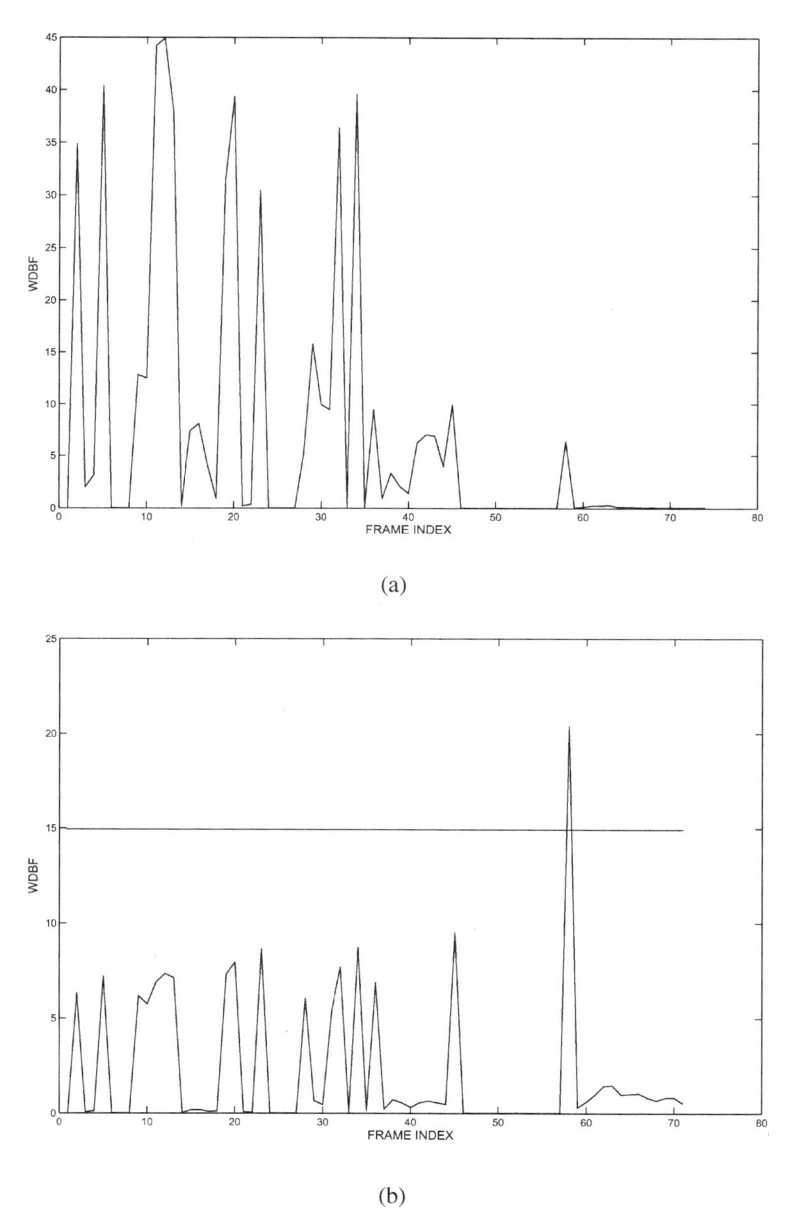

Figure 4.9 (a) Wavelet difference between frames without the proposed algorithm for the video clip (shown in Figure 4.8). (b) Wavelet difference between frames with the proposed algorithm for the video clip (shown in Figure 4.8) (true shot boundary at frame number 59)

5

Shot Boundary Detection in the Presence of Fire Flicker and Explosion

5.1 Introduction

In Chapter 4, we proposed an algorithm for shot boundary detection in the presence of flashlight. In this chapter, we will discuss the issue of shot boundary detection in the presence of fire flicker and explosion.

Abrupt illumination change due to a flashing light or Fire Flicker and Explosion (FFE) are often mistaken as a shot boundaries. Many researchers have developed methods to eliminate disturbances caused by a flashing light, but only a few methods are available for shot boundary detection under fire flicker and explosion. FFE effects occur more often than other special effects in thriller movies. The existence of FFE changes the luminance and chrominance abruptly due to sudden visual effects across the video sequence. This effect usually leads to false detection of a shot boundary. Hence, there is a need to eliminate disturbances caused by FFE or to define a suitable metric insensitive to such effects.

This chapter describes an effective method based on cross correlation coefficients, stationary wavelet transform, and combination of local and adaptive thresholds for shot boundary detection under FFE.

The rest of the chapter is organized as follows. Section 5.2 gives a detailed description of the major metrics used for shot boundary detection. Section 5.3 describes the proposed method in detail. The effectiveness of the method is validated by experiments on movie data in Section 5.4. The demonstration of results under various conditions of Fire Flicker and Explosion is described in Section 5.5.

5.2 Shot Boundary Detection Metrics Tested under Fire Flicker and Explosion

Detection of fire in a video for fire alarm systems has been studied by many researchers [27–31] but detection of shot boundaries under FFE is one of the under studied areas.

This section describes several shot boundary detection metrics proposed by different researchers such as traditional shot boundary detection metrics [3, 5, 38, 39], cross correlation, a metric for shot boundary detection in the presence of flicker [40], and a shot boundary detection metric in the presence of flashlight [2].

To analyze these metrics, we consider test sequences from three movies. Due to memory constraints, it is not possible to apply the algorithm to the entire video. Hence, we divide the video sequence into a small video clip of 200 frames and then apply the algorithm to each of these small video clips in order to determine the shot boundaries.

The mathematical symbols employed to describe these metrics are summarized as follows. Let f_i and f_{i+1} are the consecutive frames, μ_i and μ_{i+1} are the mean intensity value of these frames, σ_i and σ_{i+1} are the standard deviations of intensity value of frames f_i and f_{i+1} respectively, N is the total number of frames in a one video clip and $1 \leq i \leq N-1$, $P \times Q$ is the size of the image where $1 \leq x \leq P$, and $1 \leq y \leq Q$.

Nagasaka and Tanka [5] compared several simple statistics based on grayscale and color histograms. They found the best results by splitting the image into 16 regions, using a χ^2 test on color histograms of those regions, and discarding the eight largest differences to reduce the effects of object motion and noise. The difference measure is the sum of the absolute bin-wise histogram differences of the consecutive frames. A shot boundary is declared if this difference measure exceeds a threshold.

Zhang et al. [3] compared pixel differences, statistical differences, and several different histogram methods and found that the histogram methods were a good trade-off between accuracy and speed. A 64 bin grayscale histogram was computed over each image. If the histogram difference between consecutive frames exceeds a threshold, a cut was declared.

To make the detection of camera breaks more robust, instead of comparing individual pixels we can compare corresponding regions (blocks) in two successive frames on the basis of second-order statistical characteristics of their intensity values. One such metric for comparing corresponding regions is called the likelihood ratio [38]. The following formula computes the likelihood ratio and determine whether or not it exceeds a given threshold:

$$\frac{[(\frac{\sigma_i+\sigma_{i+1}}{2}) + (\frac{\mu_i-\mu_{i+1}}{2})^2]^2}{\sigma_i \times \sigma_{i+1}} \tag{5.1}$$

where μ_i and μ_{i+1} denote the mean intensity values for a given region in two consecutive frames, and σ_i and σ_{i+1} denote the corresponding variances. Shot boundary can now be detected by first partitioning the frame into a set of sample area. Then a shot boundary can be declared whenever the total number of sample area whose likelihood ratio exceeds the threshold is sufficiently large. This method is reasonably tolerant to noise, but is relatively slow due to the complexity of the statistical formulas. Limitation of the likelihood ratio is that if the two frames to be compared have the same mean and variance, but completely different probability density functions, no change will be detected.

Sethi and Patel [39] proposed use of the Kolmogorov Smirnov test for shot detection. If the maximum absolute value difference between cumulative distribution functions of consecutive frames exceeds a threshold, a shot boundary is detected.

The cross correlation coefficient has been widely used as a metric for shot boundary detection. The correlation between consecutive frames is computed as

$$CC(i) = \frac{\sum_{x,y}[f_i(x, y) - \mu_i] \times [f_{i+1}(x, y) - \mu_{i+1}]}{\sqrt{\sum_{x,y}[f_i(x, y) - \mu_i]^2 \times \sum_{x,y}[f_{i+1}(x, y) - \mu_{i+1}]^2}} \tag{5.2}$$

where CC is the cross correlation coefficient between consecutive frames. Then CC is found out for $1 \leq i \leq N - 1$ (i.e. for all consecutive frames in the video clip) by using Eq. (5.2) and then a threshold is applied to detect shot boundaries. A high correlation signifies similar frames, probably belonging to the same shot, whereas a low value is an indication of a shot break.

Alboil et al. [40] proposed a technique to detect abrupt transitions, when random brightness variations are present in the scene. They found out a quantity called modified sign MS of a frame f_i as

$$MS(f_i(x, y)) = \begin{cases} 1 & \text{if } f_i(x, y) > (\mu_i + T) \\ -1 & \text{if } f_i(x, y) < (\mu_i - T) \\ 0 & \text{otherwise} \end{cases} \tag{5.3}$$

for $1 \leq x \leq P$, and $1 \leq y \leq Q$, where T is a threshold. Modified sign of all frames for $1 \leq i \leq N$ is found out by using Eq. (5.3) and then a second metric is defined to see if there is any match in the modified sign of two consecutive frames. This metric is denoted as a correlation of modified sign (CMS) between two consecutive frames and is defined as

$$CMS(i) = \sum_{x,y} MS_{f_i}(x, y) - MS_{f_{i+1}}(x, y) \tag{5.4}$$

Then CMS is found out for successive consecutive frames in the video clip for $1 \leq i \leq N - 1$ using Eq. (5.4) and now another threshold is applied to this CMS vector to detect the shot boundaries.

Cheol et al. [2] proposed a robust scene change detection algorithm which can detect the scene change correctly by skipping the period of a flashing light. A local color histogram was used to extract the frame difference of consecutive frames. This frame difference is then dynamically compressed by logarithmic transform, and then four thresholds are used to detect shot boundaries.

Histogram differences are considered to be more insensitive to object motion than pixel wise differences, but they are relatively sensitive to camera motion and lighting changes. Statistic based and pixel based metrics are also sensitive to these illumination changes.

5.3 Proposed Algorithm for Shot Boundary Detection in the Presence of Fire Flicker and Explosion

An effective algorithm for shot boundary detection under FFE has been proposed, as reported in [97] and described in three stages. In the first stage, cross correlation coefficients between consecutive frames are computed. The second stage involves applying a stationary wavelet transform (SWT) to these cross correlation coefficients to avoid false positives due to FFE. Finally, in the last stage a combination of local and adaptive thresholds has been applied to find shot boundaries.

5.3.1 Computing the Cross Correlation Coefficient between Consecutive Frames

For illustrating our method, we considered a small video clip of 91 consecutive frames from the movie The Marine. In this clip sudden visual effects due to fire is present in addition to shot boundaries. Each frame is converted from an RGB to a YUV color space, and only luminance component is used to obtain grayscale frame for further analysis. The similarity between first

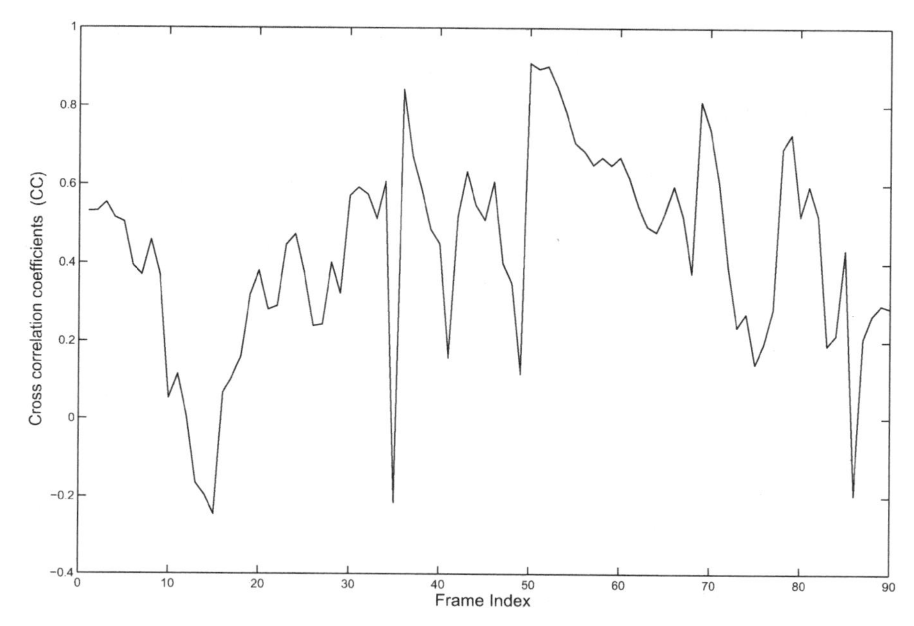

Figure 5.1 Cross correlation coefficients between consecutive frames (true shot boundaries at frame numbers 35, 49, 68, 86)

two consecutive frames is computed by using the cross correlation metric described in Eq. (5.2) and then subsequently cross correlation coefficients are obtained for all 91 consecutive frames. The cross correlation coefficients between 91 consecutive frames are denoted by 'CC' and are shown in Figure 5.1.

Here, actual shot boundaries are located at the 35th, 49th, 68th, and 86th frames. The low value of cross correlation from frames 8 to 20 is due to sudden visual effect and camera motion in these frames. From Figure 5.1 it can be seen that, if we want to detect all shot boundaries including the 68th frame, then threshold must be placed at 0.4. Any value of CC below 0.4 will be considered as a shot boundary. However this will result in many false positives, whereas if a threshold is placed at −0.1, most of the false positives could be avoided, but will result in missed shot boundaries at the 49th, 68th, and 86th frames. Hence, it is required to process cross correlation coefficients by some method, which will suppress false positives due to FFE and then automatic thresholds can be used to find out shot boundaries.

5.3.2 Applying Stationary Wavelet Transform to the Cross Correlation Coefficients

Wavelet transforms have a wide range of applications, from signal analysis to image or data compression.

One of the main concepts of wavelet theory is the interpretation of wavelet transform in terms of multiresolution decomposition. Mallat [98] described a mathematical model for the computation and interpretation of the concept of a multiresolution representation. One dimensional, three level decomposition using discrete wavelet transform [98] is shown in Figure 5.2(a).

The input signal $x(n)$ is decomposed into approximate and detail coefficients using a set of low pass (H) and high pass (G) filters followed by a decimator. These filters are quadrature mirror filters and are related by

$$h(n) = (-1)^n g(M - 1 - n)$$

where M is the filter length.

The stationary wavelet transform (SWT) is a wavelet transform algorithm designed to overcome the lack of translation-invariance of the discrete wavelet transform (DWT). The translation invariance of SWT makes it preferable in statistical application such as regression, curve estimation, spectral analysis and biomedical signal analysis [99].

In this chapter we explore the SWT for shot boundary detection. Figure 5.2(b) shows the computation of SWT algorithm. As similar to DWT, in SWT we pass the input through low pass (H) and High pass (G) filters to get coefficients C_{J-1} and D_{J-1} respectively. Now unlike DWT we do not decimate these coefficients, instead we pass them through another set of low and high pass filters which are interpolated version of H and G as shown in Figure 5.2(c).

The cross correlation coefficients (CC) obtained in Section 5.3.1 are applied as an input to the stationary wavelet transform, up to first level of decomposition, producing two sets of coefficients: approximation coefficients C_{J-1} and detail coefficients D_{J-1} using Figure 5.2(b). These coefficients are obtained by convolving cross correlation coefficients (CC) obtained from previous step with low pass filter H_j for approximation, and with the high pass filter G_j for detail. We use Daubechies second order low pass and high pass filters, $H = [-0.1294, 0.2241, 0.8365, 0.4830]$ and $G = [-0.4830, 0.8365, -0.2241, -0.1294]$ for decomposition. These kinds of wavelets [95] are used, as they are compactly supported orthonormal wavelets and they are very useful for data analysis [100]. Figure 5.3 shows

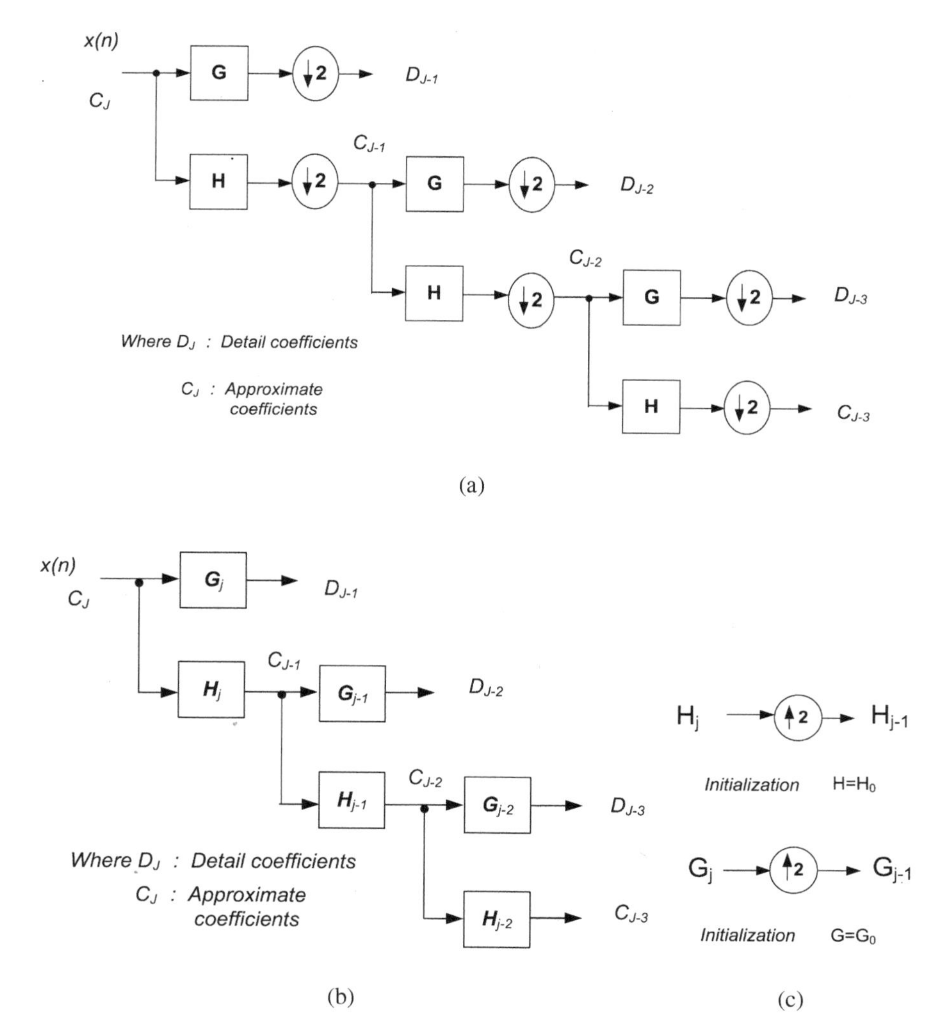

Figure 5.2 (a) Computation of DWT. (b) Computation of SWT. (c) Filters in SWT are modified at each level by upsampling

the approximation coefficients (AC) and Figure 5.4 shows detail coefficients (DC) obtained after applying SWT to cross correlation coefficients.

From Figure 5.4 it is clear that shot boundaries are clearly visible in detail coefficients and low values of CC from frame 8 to 20 as obtained in Figure 5.1 due to sudden visual effects and camera motion, is suppressed

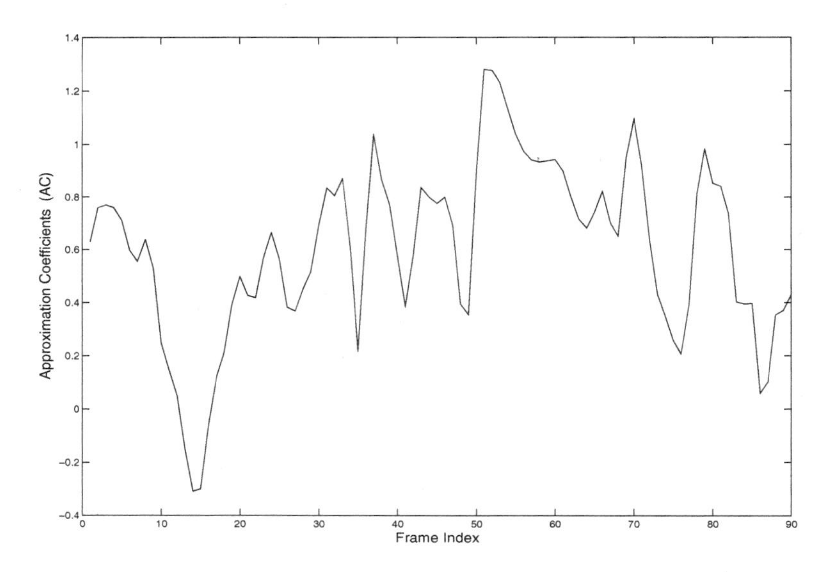

Figure 5.3 Approximation coefficients AC (true shot boundaries are at frame numbers 35, 49, 68, 86)

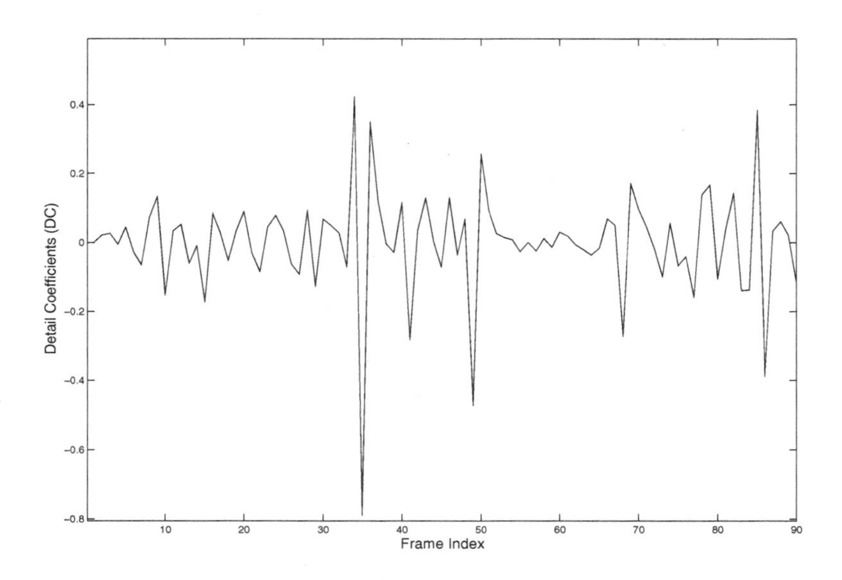

Figure 5.4 Detail coefficients DC (true shot boundaries are at frame numbers 35, 49, 68, 86)

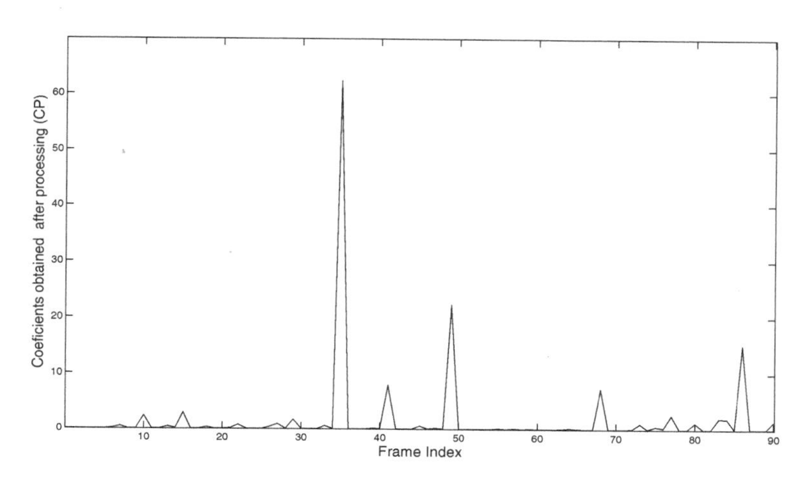

Figure 5.5 Coefficients obtained after processing detail coefficient (CP) (true shot boundaries are at frame numbers 35, 49, 68, 86)

here. Suppression of these effects using stationary wavelet transform reduces false positives due to FFE.

The detail coefficients are processed further as per the following algorithm to obtain the coefficients denoted as CP:

$$
\begin{aligned}
&\text{for } 1 \leq i \leq N-1 \\
&\{ \\
&\quad \text{If } (DC(i) > 0) \\
&\quad \{CP(i) = 0\} \\
&\quad \text{else} \\
&\quad \{CP(i) = \gamma \times |DC(i)|\} \\
&\}
\end{aligned}
$$

where γ is a scaling coefficient and used to ease the thresholding operation. The results obtained after processing detail coefficients by the above process are shown in Figure 5.5.

5.3.3 Applying Local and Adaptive Thresholds

Threshold selection is a key issue in the shot boundary detection. Global thresholds is a fixed value used for all the test video sequences.

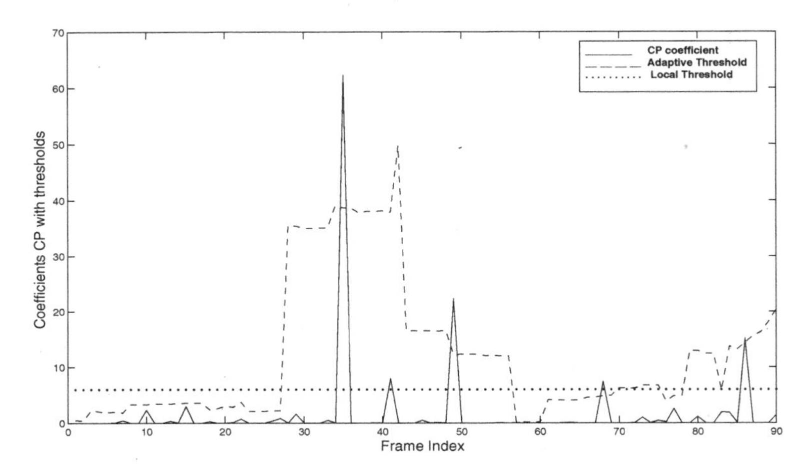

Figure 5.6 CP value with local and adaptive threshold (true shot boundaries are at frame numbers 35, 49, 68, 86)

In order to achieve a high accuracy in shot boundary detection, an appropriate threshold must be chosen [3]. As the video contents could change dramatically in the consecutive frames of the movie, it is impossible to find a single global threshold that works with all kinds of video material. Hence a combination of local and adaptive thresholds has been applied for shot boundary detection.

As the proposed algorithm is applied to an initial video clip of 200 frames and then subsequently applied to the next video clip of 200 frames till all the frames in the movie are tested by using the proposed algorithm. Hence we find the threshold locally for initial 200 frames (N frames), denoted as local threshold L_T and defined as

$$L_T = \alpha \times \frac{1}{N-1} \sum_{i=1}^{N-1} CP(i) \qquad (5.5)$$

where α is a scaling coefficient.

$$A_T(i) = \beta \times AI(i) \qquad (5.6)$$

where

$$AI(i) = \begin{cases} \frac{1}{6}\sum_{k=i+1}^{i+6} CP(k) & \text{for } 1 \leq i \leq 6 \\ \frac{1}{2}\{\frac{1}{6}\sum_{k=i-6}^{i-1} CP(k) + \frac{1}{6}\sum_{k=i+1}^{i+6} CP(k)\} & \text{for } 7 \leq i \leq N-7 \\ \frac{1}{6}\sum_{k=i-6}^{i-1} CP(k) & \text{for } N-6 \leq i \leq N-1 \end{cases}$$

As L_T depends on the mean value of 200 frames and will definitely change if the next 200 frames will be considered for analysis, this threshold is termed a local threshold.

As the local threshold is constant for 200 frames and might result in detecting false positives, an adaptive threshold has been defined which will adaptively adjust threshold for each frame using the average CP value of 6 frames preceding the current frame and the average CP value of 6 frames after the current frame. This is denoted by A_T and defined by Eq. (5.6), where β is a scaling coefficient, and AI is the average between the CP coefficients in the left sliding window preceding the CP value of a current frame, and the CP coefficients in the right sliding window after the CP value of a current frame. We have observed that the shot boundaries can be separated as close as 6 frames and therefore, we use the average CP value of 6 frames preceding the current frame and the average CP value of 6 frames after the current frame to calculate the adaptive threshold. A smaller window size results in higher false positives and a larger window size would result in higher missed detection.

AI is calculated by the following method. First we put the average CP value of 6 frames preceding the current frame into left sliding window. Simultaneously the average CP value of 6 frames after the current frame is put into right sliding window. Then we use the average CP of all frames within right and left sliding window to calculate AI. The advantages of using the above process to calculate AI lies in two facts: (1) the FFE effects always make the average CP value of frame large, and (2) there are several frames where visual change due to FFE is significant within the right or left sliding window. For the first 6 frames the left sliding window does not exist, hence the average CP value of the right sliding window is considered, whereas for the last 6 frames the right sliding window does not exist, hence the average CP value of a left sliding window is considered. Figure 5.6 shows local and adaptive thresholds applied to CP, where a local threshold is denoted by a dotted line and an adaptive threshold is denoted by a dashed line. If the value of CP at the ith frame is greater than both, L_T at the ith frame and A_T at the ith frame, then shot boundary is declared. From Figure 5.6 it is seen that

Table 5.1 Number of shot boundaries, and number of frames with Fire Flicker and Explosion effects used in each movie for the analysis

Movie	PH	TM	SPR
Shot boundaries	65	198	170
Frames with FFE	1936	3575	3748

frame numbers 35, 49, 68, and 86 satisfy the above condition and hence it is declared as a correct detection.

5.4 Experimental Results and Discussion

5.4.1 Test Video Sequence and an Evaluation Criterion

The proposed algorithm has been tested on various movies such as Pearl Harbor (PH), The Marine (TM) and Saving Private Ryan (SPR). These movies are manually observed frame by frame to find actual shot boundaries. These movies are considered for obtaining test data since FFE scenes are found before, after and during shot boundaries. Number of shot boundaries and number of frames with FFE effects for each movie is shown in Table 5.1. The test data also contain frames with camera and object motion before or after shot boundaries.

We used Recall, Precision and F1 measure as discussed in Section 2.5 as an evaluation metric to compare shot boundary detection algorithms.

5.4.2 Performance Comparison of the Proposed Algorithm with Major Shot Boundary Detection Metrics

The performance of the proposed algorithm is compared with traditional shot boundary detection metrics [3, 5, 38, 39], and cross correlation (these metrics have not taken illumination changes into account), a metric for shot boundary detection in the presence of flicker [40], and a shot boundary detection metric in the presence of flashlight [2].

We empirically determine the values of the scaling parameters based on our investigation with the movie Pearl Harbor, and then use the same values for our investigation with other movies without any further tuning of these parameters. The scaling parameter γ, α (Eq. 5.5), and β (Eq. 5.6) were empirically chosen to be 10, 3, and 4, respectively. The performance comparison of different metrics with the proposed algorithm has been shown in Tables 5.2, 5.3, and 5.4 for the movies Pearl Harbor, The Marine and Saving Private Ryan, respectively.

Table 5.2 Performance comparison of the proposed algorithm for the movie Pearl Harbor

Algorithm	D	C	M	FP	R	P	F1
Nagasaka [5]	65	30	35	38	46.15	44.12	45.11
Zhang et al. [3]	65	35	30	55	53.84	38.88	45.15
Jain et al. [38]	65	41	24	75	63.08	35.34	45.30
Sethi et al. [39]	65	59	06	85	90.77	40.94	56.42
Cross correlation	65	60	05	60	92.30	50	64.86
Albiol et al. [40]	65	59	06	28	89.39	67.82	77.12
Cheol et al. [2]	65	56	09	16	86.15	77.78	81.75
Proposed	65	65	00	05	100	92.86	96.29

Table 5.3 Performance comparison of the proposed algorithm for the movie The Marine

Algorithm	D	C	M	FP	R	P	F1
Nagasaka [5]	198	41	157	65	20.71	38.68	26.97
Zhang et al. [3]	198	76	122	121	38.38	38.57	38.47
Jain et al. [38]	198	114	84	229	57.58	33.24	42.14
Sethi et al. [39]	198	73	125	115	36.87	38.83	37.82
Cross correlation	198	134	64	63	67.67	68.02	67.84
Albiol et al. [40]	198	92	106	144	46.46	38.98	42.39
Cheol et al. [2]	198	159	39	290	80.30	35.41	49.14
Proposed	198	176	22	23	88.89	88.44	88.66

Table 5.4 Performance comparison of the proposed algorithm for the movie Saving Private Ryan

Algorithm	D	C	M	FP	R	P	F1
Nagasaka [5]	170	52	118	264	30.59	16.46	21.40
Zhang et al. [3]	170	71	199	466	41.76	13.22	20.08
Jain et al. [38]	170	69	101	82	40.59	45.70	42.99
Sethi et al. [39]	170	41	129	108	24.12	27.52	25.70
Cross Correlation	170	116	54	47	68.24	71.17	69.67
Albiol et al. [40]	170	48	122	133	28.24	26.52	27.35
Cheol et al. [2]	170	130	40	37	76.47	77.84	77.14
Proposed	170	163	07	31	95.88	84.02	89.55

These compared metrics falsely detected sudden visual changes which occur in the consecutive frames due to FFE effects as a shot boundary in the following situations:

- When the flicker is present at different direction and position in the consecutive frames (Figure 5.7(a)).
- When both fire and flicker are present in the consecutive frames (Figure 5.7(b)).

Figure 5.7 (a) Flicker is present at different positions and directions. (b) Combination of fire and flicker. (c) Change in area and position of fire. (d) Explosion which gradually increases from small portion to full frame. (e) Object with fire moves from one direction to the other

- When the area and position of fire changes in the consecutive frames (Figure 5.7(c)).
- When the explosion gradually increases from a small portion in the frame to the full frame in the consecutive frames (Figure 5.7(d)).
- When the object with fire moves from one position to the other in consecutive frames as shown in Figure 5.7(e).

Traditional shot boundary detection algorithms [3, 5, 38, 39] are sensitive to disturbances caused by camera motion, object motion and sudden illumination changes, and produces false positives for such cases. Algorithm proposed by Albiol et al. [40] for shot boundary detection in the presence of flicker is a pixel difference based method, and is still sensitive to the disturbances caused by FFE. The method suggested by Cheol et al. [2] for shot boundary detection in the presence of flashlight does not take into account sudden visual changes due to FFE effects and failed in such cases. Color ratio histogram [2] performs relatively better than other tested metrics for shot boundary detection and is also robust to small object motion. The false positives detected by this algorithm are due to the disturbances caused by FFE and camera motion. Cross correlation based algorithms also perform better than the other tested metrics and the F1 measure is found to be very close to that of color ratio histogram. This algorithm detects false positives due to fast camera and object motion, but is more robust to FFE.

The proposed algorithm is found to achieve a high F1 measure compared to other tested algorithms. The false positives detected by our proposed algorithm are due to the disturbances caused by fast camera and object motion.

5.5 Demonstration of Results under Fire Flicker and Explosion Using the Proposed Algorithm

5.5.1 When Smoke and Explosion Is Present in the Consecutive Frames

Figure 5.8 shows frames from the movie Pearl Harbor where smoke and explosion can be observed in the consecutive frames which gradually increase from small portion of the frame to full frame. Here shot boundaries are at frame numbers 13, 28, 65, 84, and 106. Figure 5.9(a) shows results obtained after applying the cross correlation metric to the above video clip. If the global threshold is applied to this frame difference to detect all shot boundaries mentioned above, then a large number of false positives will be

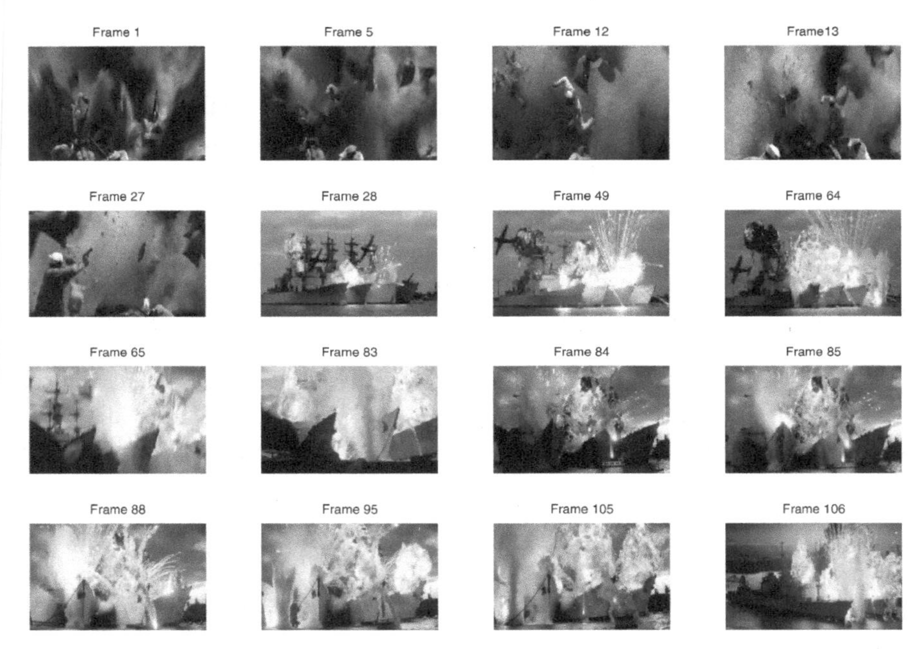

Figure 5.8 Video clip from the movie Pearl Harbor when smoke and explosion is present in the consecutive frames

detected due to smoke and explosion. Figure 5.9(b) shows results after applying the proposed algorithm. Coefficients of CP greater than the local threshold (shown by dotted lines) and an adaptive threshold (shown by dashed lines) are declared as a shot boundaries. It is clearly observed from Figure 5.9(b) that all shot boundaries are correctly detected by the proposed algorithm though visual difference between consecutive frames due to smoke and explosion is large as shown in Figure 5.8.

5.5.2 When a Combination of Fire Flicker and Explosion Is Present in the Consecutive Frames

Figure 5.10 shows frames from the movie The Marine where fire flicker and explosion can be observed in the consecutive frames at different direction and position. Here shot boundaries are at frame number 7, 45, 64, 74, and 89. Figure 5.11(a) shows results obtained after applying cross correlation metric to the above video clip. If the global threshold is applied to this frame difference to detect all shot boundaries including shot boundary at frame

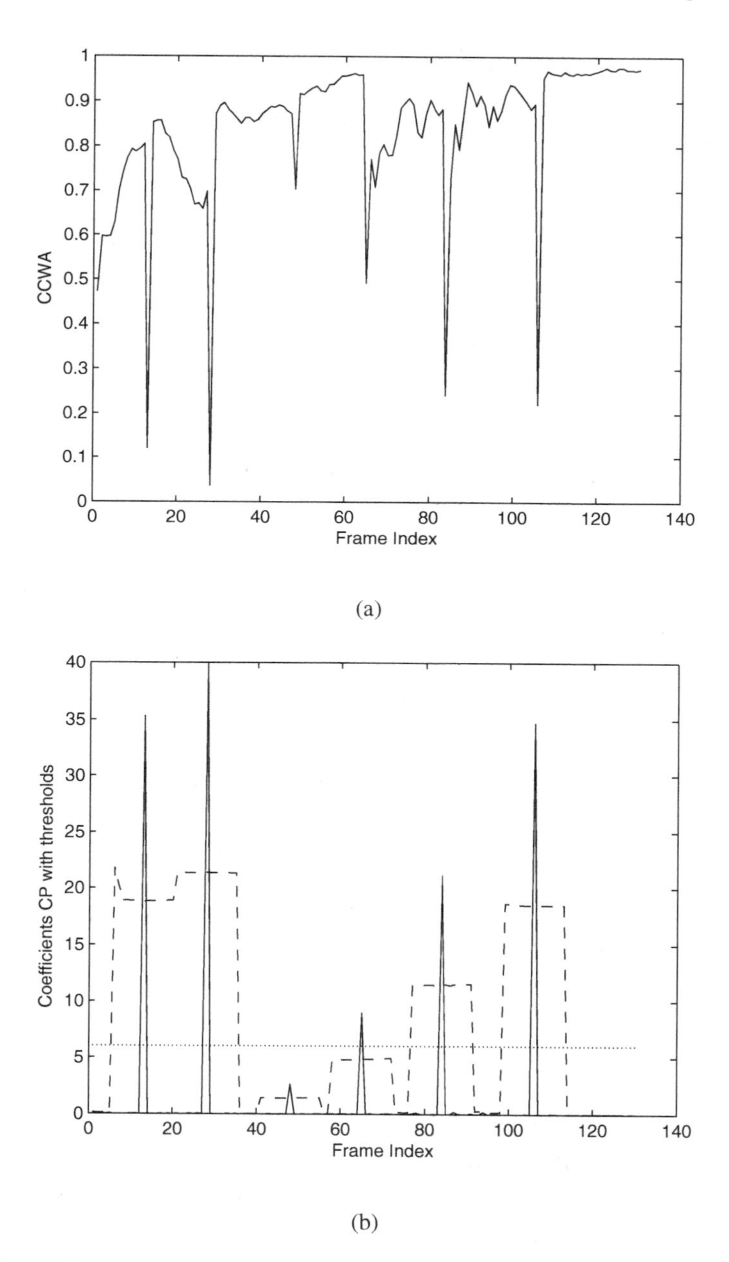

Figure 5.9 (a) Result using the cross correlation metric for the video clip (shown in Figure 5.8). (b) Result using the proposed algorithm for the video clip (shown in Figure 5.8) (true shot boundaries are at frame numbers 13, 28, 65, 84, 106)

Figure 5.10 Video clip from the movie The Marine when the combination of Fire Flicker and Explosion is present in the consecutive frames

number 7, then false positives detected will be more than actual shot boundaries. Figure 5.11(b) shows the results after applying the proposed algorithm. Coefficients of CP greater than the local threshold (shown by dotted lines) and an adaptive threshold (shown by dashed lines) are declared as a shot boundaries. It is clearly observed from Figure 5.11(b) that all shot boundaries are correctly detected by the proposed algorithm. Cross correlation coefficient metric produce more false positives due to Fire Flicker and Explosion as observed in Figure 5.11(a), whereas the proposed algorithm have successfully detected all shot boundaries without any false positives.

5.5.3 When the Area and Position of Fire in the Consecutive Frames Is Changed

Figure 5.12 shows frames from the movie The Marine where a change in area and position of fire in the consecutive frames is observed, and also the object with fire moves from one direction to other direction in the consecutive frames. Here shot boundaries are at frame numbers 15, 39, 66 and 89. Fig-

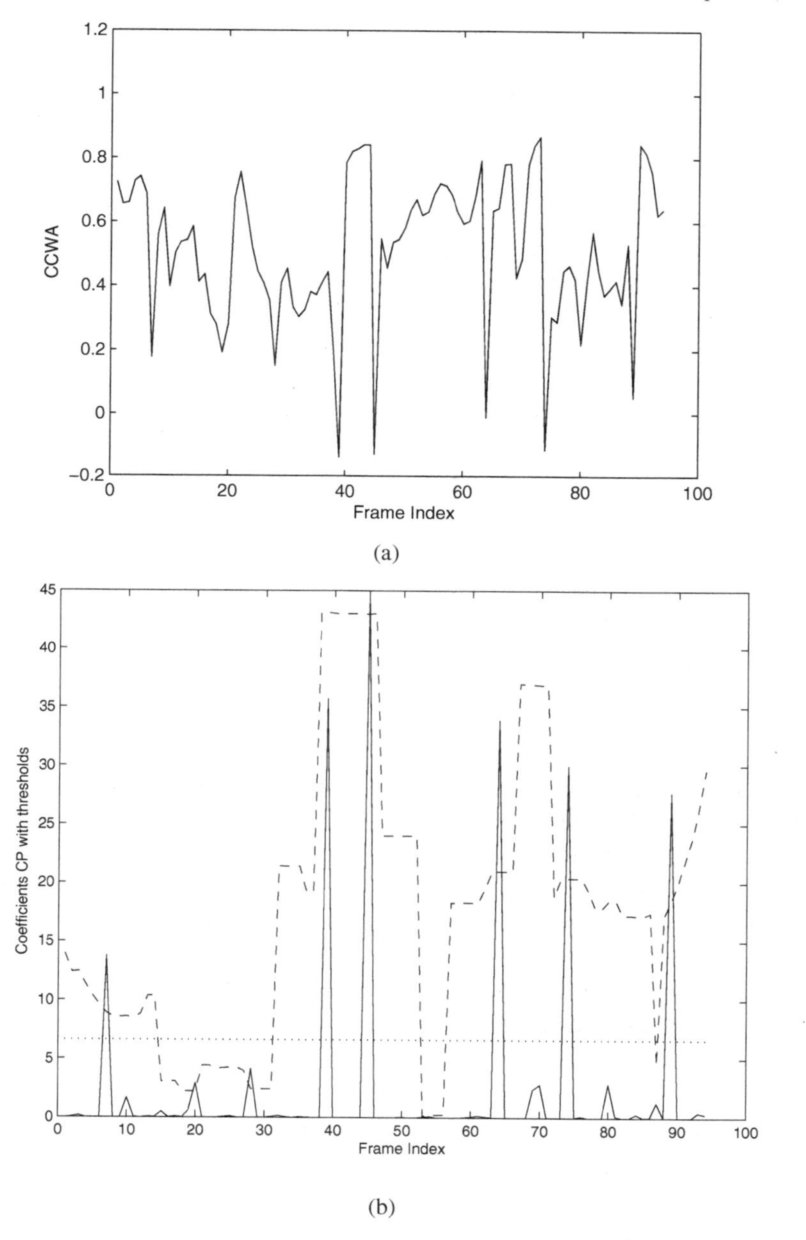

Figure 5.11 (a) Result using the cross correlation metric for the video clip (shown in Figure 5.10). (b) Result using the proposed algorithm for the video clip (shown in Figure 5.10) (true shot boundaries are at frame numbers 7, 45, 64, 74, 89)

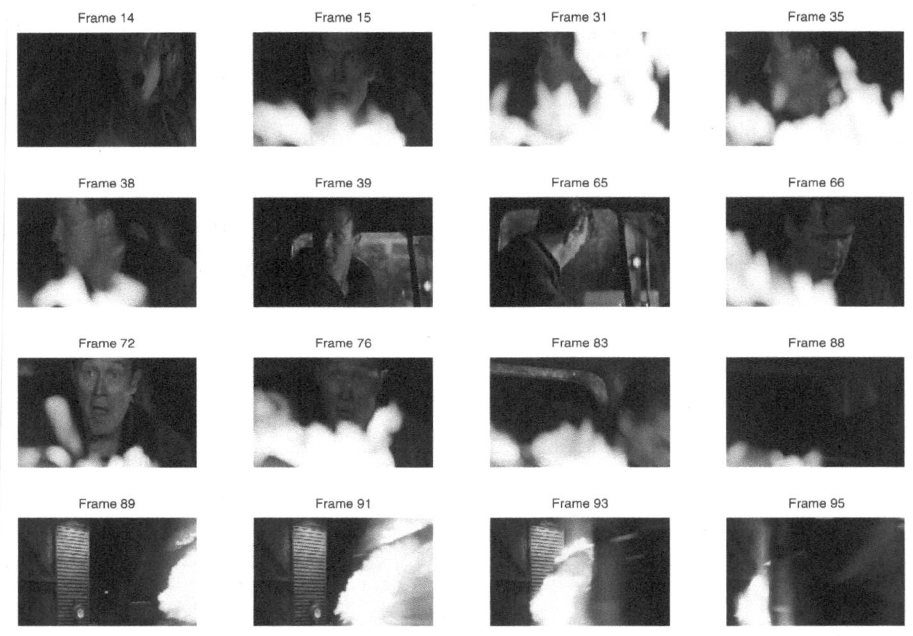

Figure 5.12 Video clip from the movie The Marine when the area and position of fire in the consecutive frames is changed

ure 5.13(a) shows the results obtained after applying the cross correlation metric to this video clip. If the global threshold is applied to this frame difference to detect all shot boundaries mentioned above, then some false positives will be detected. Figure 5.13(b) shows the results after applying the proposed algorithm. Coefficients of CP greater than the local threshold (shown by dotted lines) and an adaptive threshold (shown by dashed lines) are declared as a shot boundaries. It is clearly observed from Figure 5.13(b) that all shot boundaries are correctly detected by the proposed algorithm in the presence of fire whereas other algorithms produces false positives.

5.5.4 When the Object with Fire Moves in Addition to a Change in Area and Position of Fire

Figure 5.14 shows frames from the movie The Marine where camera motion with fire is observed from frames 1–30, change in area and position of fire from frames 35–48, object with fire moves from one direction to other direction from frames 49–67, and again change in area and position

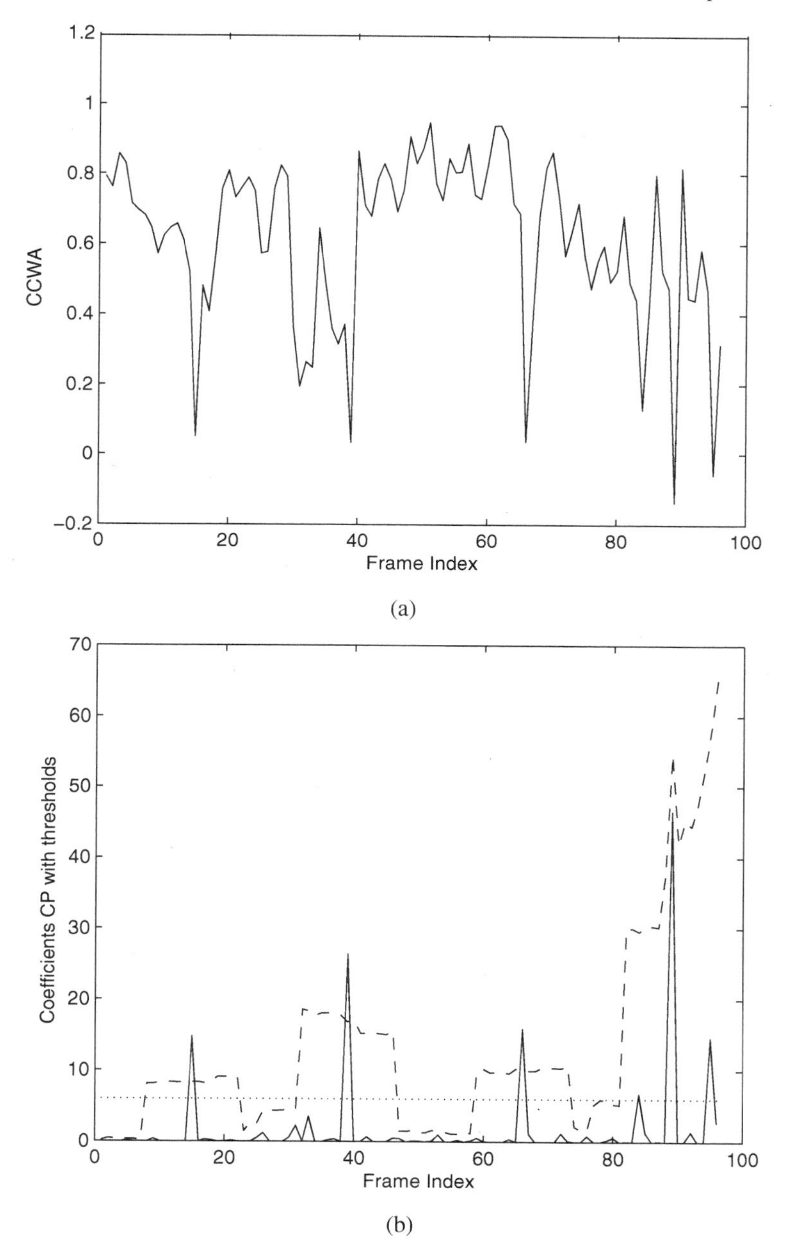

Figure 5.13 (a) Result using the cross correlation metric for the video clip (shown in Figure 5.12). (b) Result using the proposed algorithm for the video clip (shown in Figure 5.12) (true shot boundaries are at frame numbers 15, 39, 66, 89)

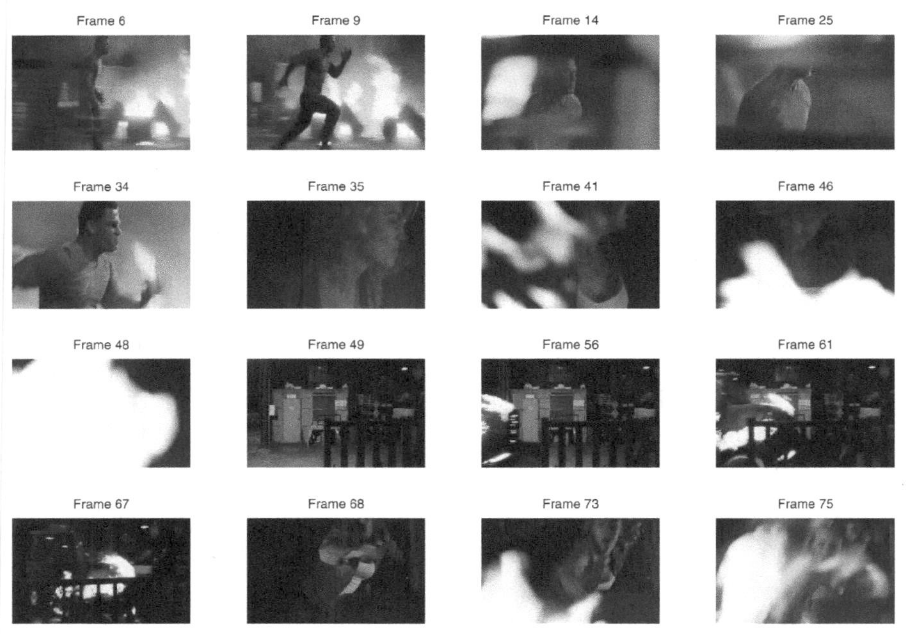

Figure 5.14 Video clip from the movie The Marine when the object with fire moves in addition to change in area and position of fire

of fire is observed from frame 66 onwards. Here shot boundaries are at frame numbers 35, 49, 68, and 86. Figure 5.15(a) shows the results obtained after applying the cross correlation metric to the above video clip. If the global threshold is applied to this frame difference to detect all shot boundaries including shot boundary at frame number 68, then false positives detected will be more than actual shot boundaries. Figure 5.15(b) shows the results after applying the proposed algorithm. Coefficients of CP greater than the local threshold (shown by dotted lines) and an adaptive threshold (shown by dashed lines) are declared as a shot boundaries. It is clearly observed from Figure 5.15(b) that all the shot boundaries are correctly detected by the proposed algorithm. Cross correlation coefficient metric produce more false positives due to camera motion, and a change in area and position of fire as observed in Figure 5.15(a), whereas the proposed algorithm has successfully detected all shot boundaries without any false positives.

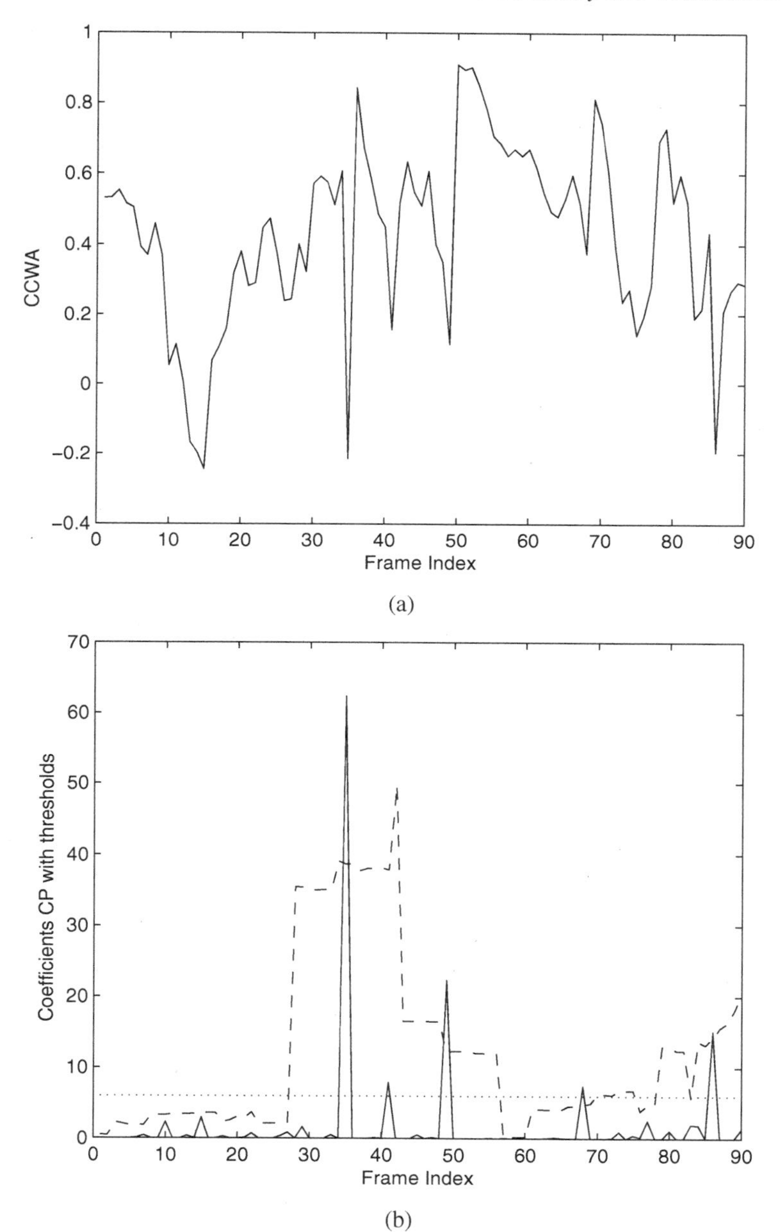

Figure 5.15 (a) Result using the cross correlation metric for the video clip (shown in Figure 5.14). (b) Result using the proposed algorithm for the video clip (shown in Figure 5.14) (true shot boundaries are at frame numbers 35, 49, 68, 86)

5.6 Summary and Conclusions

Detection of shot boundaries in the thriller movies under FFE effects is a difficult task. The major problem arises from the fact that changes in luminance is not uniform over consecutive frames during this effect, as the direction and position of fire moves in the consecutive frames. In a few cases, appearance and disappearance of fire in the consecutive frames contributes to large differences. In this chapter we have evaluated the effectiveness of various metrics under FFE effects. It has been found that, color ratio histogram and cross correlation coefficient metric performs better than the other compared metrics under these effects. The behavior of FFE under which these metrics failed has been discussed. We have proposed an effective algorithm based on cross correlation coefficient, stationary wavelet transform and combination of local and adaptive thresholds for shot boundary detection. Experimental results on movie data show significant improvements in terms of better Recall and Precision.

6

Shot Boundary Detection in the Presence of Illumination Variation and Motion

6.1 Introduction and Related Work

In Chapters 4 and 5, we proposed an algorithm for shot boundary detection in the presence of flashlight, fire flicker and explosion respectively. The objective of this chapter is to investigate the performance of major metrics for shot boundary detection specifically in the presence of camera and object motion. For comparing different metrics we consider the video clips as the test video sequence, where fast camera and object motion is observed in addition to shot boundaries. The results of this investigation motivate the algorithm proposed in Section 6.3.

The elimination of disturbances due to illumination and motion is a major challenges to the shot boundary detection algorithms. Many researchers have reported the method for avoiding false alarms due to illumination variation during detection of shot boundary, see for example [2, 4, 24–26, 58–60], and also due to the influence of camera and object motion in the video segmentation [33–37]. It is a challenging task to develop a single approach which will solve both issues and hence we propose a method for shot boundary detection in the presence of illumination and motion.

In this proposed algorithm, extraction of structure feature of each frames is carried out using dual tree complex wavelet transform, followed by detection of shot boundaries by applying spatial domain structure similarity algorithm to the structure feature of consecutive frames. Finally correct shot boundaries are declared by using local and adaptive threshold.

The rest of the chapter is organized as follows. In Section 6.2 performance evaluation of various metrics in the presence of motion is discussed. The evaluation results of these traditional metrics in RGB, HSV and YUV color spaces are also discussed. In Section 6.3, we propose an algorithm using dual tree complex wavelet transform and spatial domain structural similarity algorithm for shot boundary detection in the presence of illumination and motion. Performance comparison of the proposed algorithm with other shot boundary detection methods is also described.

6.2 Performance Evaluation of Various Metrics in the Presence of Motion

6.2.1 Major Metric Used for Evaluation

The mathematical symbols employed to describe these metrics are summarized as follows. Let f_i and f_{i+1} be the consecutive frames, μ_i and μ_{i+1} are the mean intensity value of these frames, σ_i and σ_{i+1} are the standard deviations of intensity value of frames f_i and f_{i+1} respectively, N is the total number of frames in one video clip and $1 \leq i \leq N - 1$, $P \times Q$ is the size of the image where $1 \leq x \leq P$, and $1 \leq y \leq Q$.

Pixel Differences

The simplest approach to detect, if the two images are significantly different, is to count the number of pixels that have changed. The pixel differences (denoted as PD) metric is defined as

$$PD(i) = \frac{1}{PQ} \sum_{x=1}^{P} \sum_{y=1}^{Q} |(f[x, y, i] - f(x, y, i + 1)| \tag{6.1}$$

This technique is very sensitive to camera and object motion. Zhang et al. [3] suggested that the effect of motion can be reduced by using 3×3 averaging filter before pixel-wise comparison.

Likelihood Ratio

Jain et al. [38] computed a likelihood ratio test based on the assumption of uniform second order statistics. This is discussed in Section 5.2.

Histogram Difference

Histograms are the most common method used to detect shot boundaries. Histogram difference is defined by

$$HD(i) = \sum_{j=1}^{G} |(H_i[j] - H_{i+1}[j])| \tag{6.2}$$

where $H_i[j]$ and $H_{i+1}[j]$ denote the histogram value for the ith and $(i+1)$th frame, respectively. j is one of the G possible gray level.

The histogram comparison algorithm is less sensitive to object motion than pixel differences. There may be certain cases in which two images have a similar histogram but completely different content, but such cases are rare in practice.

Chi-Square Test

Nagasaka and Tanka [5] experimented with histogram and pixel difference metrics, and concluded that histogram metrics are the most effective.

The chi-square test (denoted as CS) is defined as

$$CS(i) = \sum_{j=1}^{G} \frac{|(H_i[j] - H_{i+1}[j])|^2}{H_{i+1}[j]} \tag{6.3}$$

where $H_i[j]$ denote the histogram value for the ith frame and j is one of the G possible gray level.

Color Histogram

A color histogram comparison calculated by histogram comparison of each color space of adjacent two frames is defined as

$$HD_{r,g,b}(i) = \sum_{j=1}^{G} (|(H_i^r[j] - H_{i+1}^r[j])| + |(H_i^g[j] - H_{i+1}^g[j])| + |(H_i^b[j] - H_{i+1}^b[j])| \tag{6.4}$$

where $H_i^r[j]$, $H_i^g[j]$ and $H_i^b[j]$ denote the histogram value for the ith frame in R, G and B color space respectively.

Table 6.1 Number of frames considered for analysis from each test video

Movie	XM	HA	MI	JMP	WED	PR	GBAU	BEE
Number of frames	16340	11637	12000	14725	23895	26995	3264	4750

Table 6.2 Performance comparison between likelihood ratio and color histogram in RGB color space

Metric	Video→	XM	HA	MI
LHR	R	74.01	82.58	49.47
	P	58.59	86.48	19.31
	F1	65.40	84.48	27.78
CH	R	81.81	86.18	45.45
	P	73.35	93.57	43.62
	F1	77.34	89.72	44.52

6.2.2 Test Video Sequence and an Evaluation Criterion Used for Metric Evaluation

These traditional metrics are tested on movies X-Men (XM), Home Alone (HA), Mission Impossible 3 (MI), Jumper (JMP), Wednesday (WED), Pale Rider (PR), Bee movie (BEE), Good Bad and Ugly (GBU). These movies are manually observed frame by frame to find actual shot boundaries. These movies are considered for obtaining test data since large number of frames are observed with object motion and camera motion. We mostly considered the video clips where fast camera and object motion is observed in addition to shot boundaries. The number of frames considered for test video sequence in each movie is shown in Table 6.1.

We used Recall, Precision and F1 measure as discussed in Section 2.5 as an evaluation metric to compare shot boundary detection algorithm.

6.2.3 Evaluation Results of the Traditional Shot Boundary Detection Metrics in RGB, HSV and YUV Color Spaces

In this chapter, the performance of metrics such as pixel difference, histogram difference, likelihood ratio, color histogram, chi-square test in RGB, HSV and YUV color spaces are compared on the same data sequence as reported in [101]. We used a local threshold described in Section 5.3.3 to declare a shot boundary in all these metrics. The performance comparison between likelihood ratio and color histogram in RGB color space are shown in Table 6.2. It has been observed that color histogram provided better results than likelihood ratio in terms of F1 measure. False positives and missed detection in both algorithms was due to the fast camera and object motion.

Table 6.3 Performance comparison between likelihood ratio and histogram difference in HSV color space (using only the Hue component)

Metric	Video→	XM	HA	MI
LHR	R	63.07	97.08	36
	P	73.55	75.47	40.34
	F1	67.91	84.92	38.05
HD	R	54.72	63.30	54.38
	P	40.93	83.30	31.89
	F1	46.83	71.93	40.20

Table 6.4 Performance comparison between chi-square and pixel difference in YUV color space (using only the Y component)

Metric	Video→	JMP	WED	PR	GBU	BEE
CS	R	58.06	73.40	60.86	72.5	70
	P	65.85	74.22	81.41	52.70	61.7
	F1	61.71	73.80	69.65	61.03	65.58
PD	R	41	69.23	66	57.7	46.6
	P	64.02	77.80	66	76.3	48.43
	F1	49.98	73.26	66	65.70	47.42

The performance comparison between likelihood ratio and histogram difference in HSV color space (using only the Hue component) are shown in Table 6.3. We observed that the performance of the likelihood ratio is better than the histogram difference in HSV color space. The performance of likelihood ratio is slightly better in HSV color space when compared to RGB color space, whereas the performance of color histogram in RGB color space is better than grayscale histogram in HSV color space. The performance of these metrics was found poor in both the color space for the movie Mission Impossible due to the large number of frames with fast camera and object motion.

We also tested the performance of chi-square test and pixel difference method in YUV color space (using only the Y component) for various movies and the results are shown in Table 6.4. The results of chi-square were found to be better than pixel difference method in terms of F1 measure.

Overall it has been observed that these metrics did not perform well due to the disturbances caused by fast camera and object motion. The maximum false positives and missed detections were due to frame difference between consecutive frames caused by fast camera motion.

6.3 Proposed Algorithm to Detect Shot Boundaries in the Presence of Illumination and Motion

In order to evaluate discontinuity between frames based on the selected features an appropriate metric needs to be chosen. Traditional metrics used for shot boundary detection are broadly classified as histogram based metrics, statistic based metrics, pixel difference based metric, edge based metrics and MPEG metrics. Histogram and statistic based methods are sensitive to lighting changes but are invariant to changes in object motion. The pixel difference comparisons are more robust to lighting changes and are sensitive to motion and camera zooming and panning. The edge based methods are invariant to lighting changes and are generally used in combination with histogram based method, but this approach is computation wise very expensive. Hence we propose a single approach for shot boundary detection to address both the issues, i.e. illumination and motion. We used dual tree complex wavelet transform (DT-CWT) instead of traditional metrics to extract features for shot boundary detection because of its shift invariance, rotational invariance and directional selectivity property. These geometrical structure features obtained by DT-CWT are approximately invariant to motion. The structures of the objects in the scene are independent of the illumination, also it has been shown by Heng and Ngan [26] that edge features are invariant to illumination changes and several types of motion. Wang et al. [102] showed that the structural information in an image are those attributes that represent the structure of objects in the scene, independent of average luminance and contrast. Hence the structure features obtained by DT-CWT, which are edges of the objects, are invariant to illumination changes. The human visual system is highly adapted to extract structure information from the visual scene, and therefore a measurement of structural similarity should provide a good metrics to find similarity or dissimilarity between consecutive frames. Spatial domain structural similarity algorithm (SSIM) is very useful in finding similarity between frames using structure features, and is already used in image retrieval algorithms. We explore SSIM algorithm as a metric for shot boundary detection. SSIM is sensitive to translation, rotation and scaling of the images but is invariant to illumination changes. Hence initially DT-CWT can be used to find out structure features, which are insensitive to these distortion, and then SSIM algorithm can be used to find out shot boundaries. In this section, we propose an algorithm using dual tree complex wavelet transform and spatial domain structural similarity algorithm for shot boundary detection in the presence of illumination and motion, as reported in [103].

In this proposed algorithm, extraction of structure features in each frame is done using dual tree complex wavelet transform, followed by detection of shot boundaries by applying spatial domain structure similarity algorithm to the structure features of consecutive frames, and finally correct shot boundaries are declared by using local and adaptive thresholds.

6.3.1 Background of DTCWT and SSIM Index

In this section, we review dual tree complex wavelet transform and the spatial domain structural similarity algorithm (SSIM). The dual tree complex wavelet transform is used to find the structure features of each frame and the SSIM algorithm is used to find potential shot boundaries.

Dual-Tree Complex Wavelet Transform

Discrete wavelet transform (DWT) has poor directional selectivity and also lacks shift invariance. A separable 2D-DWT can be computed efficiently in discrete space by applying the associated one dimensional filter to each column of the image and then applying the filter to each of the resultant coefficients. Therefore a 2D-DWT produces four bandpass subimages at each level, corresponding to low-low, low-high, high-low, and high-high filtering. The low-low part coefficients represent the smooth version of the original function. However the other three sub-bands wavelet coefficients of two dimensional DWT capture features along lines at an angle of 0, 90 and 45°. To overcome the drawbacks of DWT (poor directional selectivity and shift variance), the dual tree complex wavelet transform (DT-CWT) has been developed [104, 105], which allows perfect reconstruction in addition to shift invariance and directional selectivity. It has the ability to differentiate positive and negative frequencies and also can decompose a signal in terms of complex shifted and dilated mother wavelet. The DT-CWT is implemented using separable transforms and by combining subband signals appropriately [106]. A complex-valued wavelet $\Psi(x)$ can obtained as

$$\Psi(x) = \Psi_h(x) + j\Psi_g(x) \tag{6.5}$$

where $\Psi_h(x)$ and $\Psi_g(x)$ are both real-valued wavelets and $\Psi(x)$ is a complex wavelet. The procedure for obtaining the 1D dual tree complex wavelet trans-

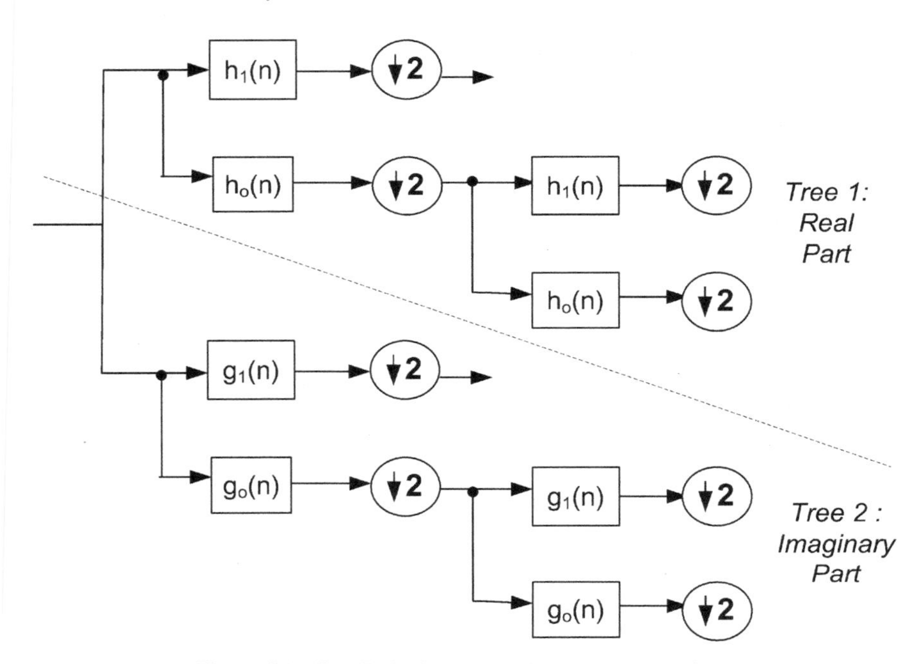

Figure 6.1 The 1D dual tree complex wavelet transform

form is shown in Figure 6.1, where two real wavelet trees are used and each tree is capable of perfect reconstruction.

One tree generates the real part of the transform and the other is used to generate a complex part [107]. As shown in Figure 6.1, $\{h_o(n), h_1(n)\}$ and $\{g_o(n), g_1(n)\}$ is a quadrature mirror filter pair in the real tree of analysis branch and complex tree of analysis branch respectively. All filter pairs are orthogonal and real valued. If the filters in both trees are made to be offset by half-sample, two wavelets satisfy Hilbert transform pair conditions [106] and an approximately analytic wavelet is given in Eq. (6.5), where $\Psi_g(x)$ is approximately the Hilbert transform of $\Psi_h(x)$.

2D Dual-Tree CWT The detail explanation for 2D dual-tree complex wavelet transform has been given in [107]. Consider the 2D wavelet $\Psi(x, y) = \Psi(x)\Psi(y)$ associated with the row-column implementation of the wavelet transform, where $\Psi(x)$ is a complex wavelet given by Eq. (6.5). Then $\Psi(x, y)$ is

$$\Psi(x, y) = [\Psi_h(x) + j\Psi_g(x)][\Psi_h(y) + j\Psi_g(y)] \tag{6.6}$$

Therefore

$$\Psi(x, y) = \Psi_h(x)\Psi_h(y) - \Psi_g(x)\Psi_g(y) + j[\Psi_g(x)\Psi_h(y) + \Psi_h(x)\Psi_g(y)] \tag{6.7}$$

If we take the real part of this complex wavelet, then the sum of two separable wavelets is obtained as

$$\text{Real Part}[\Psi(x, y)] = \Psi_h(x)\Psi_h(y) - \Psi_g(x)\Psi_g(y) \tag{6.8}$$

The support of the spectrum of this real wavelet is oriented at $-45°$. The first term in Eq. (6.8), $\Psi_h(x)\Psi_h(y)$ is the HH wavelet of a separable 2D real wavelet transform implemented using the filters $\{h_o(n), h_1(n)\}$. The second term $\Psi_g(x)\Psi_g(y)$ is also the HH wavelet of a separable 2D real wavelet transform implemented using the filters $\{g_o(n), g_1(n)\}$.

To obtain the other five oriented real 2D wavelets, the procedure can be repeated on the following complex 2D wavelets: $\Psi(x)\Psi^*(y)$, $\varphi(x)\Psi(y)$, $\Psi(x)\varphi(y)$, $\varphi(x)\Psi^*(y)$, and $\Psi(x)\varphi^*(y)$, where $\Psi(x) = \Psi_h(x) + j\Psi_g(x)$, $\varphi(x) = \varphi_h(x) + j\varphi_g(x)$, and $*$ represents the complex conjugate.

Then the following six wavelets are obtained by

$$\Psi_i(x, y) = \frac{1}{\sqrt{2}}(\Psi_{1,i}(x, y) - \Psi_{2,i}(x, y)) \tag{6.9}$$

$$\Psi_{i+3}(x, y) = \frac{1}{\sqrt{2}}(\Psi_{1,i}(x, y) + \Psi_{2,i}(x, y)) \tag{6.10}$$

for $i = 1, 2, 3$, where the two separable 2D wavelet bases are defined as

$$\Psi_{1,1}(x, y) = \varphi_h(x)\Psi_h(y), \quad \Psi_{1,2}(x, y) = \Psi_h(x)\varphi_h(y) \tag{6.11}$$

$$\Psi_{1,3}(x, y) = \Psi_h(x)\Psi_h(y), \quad \Psi_{2,1}(x, y) = \varphi_g(x)\Psi_g(y) \tag{6.12}$$

$$\Psi_{2,2}(x, y) = \Psi_g(x)\varphi_g(y), \quad \Psi_{2,3}(x, y) = \Psi_g(x)\Psi_g(y) \tag{6.13}$$

These six real parts of the complex wavelet sub-bands of the 2D DT-CWT are strongly oriented in $\{+15°, +45°, +75°, -15°, -45°, -75°\}$ directions. Thus 2D dual-tree wavelet transform are not only oriented but also approximately analytic and hence make the CWT shift invariance in nature [107].

To find the imaginary part of the complex wavelet, consider the imaginary part of Eq. (6.7)

$$\text{Imag Part}[\Psi(x, y)] = \Psi_g(x)\Psi_h(y) + \Psi_h(x)\Psi_g(y) \tag{6.14}$$

This wavelet is also oriented at $-45°$, which is the same as the spectrum of the real part in Eq. (6.8). The first term in Eq. (6.14), $\Psi_g(x)\Psi_h(y)$ is the HH wavelet of a separable 2D real wavelet transform implemented using the filters $\{g_o(n), g_1(n)\}$ on the rows, and the filters $\{h_o(n), h_1(n)\}$ on the columns of the image. The second term $\Psi_h(x)\Psi_g(y)$ is also the HH wavelet of a separable 2D real wavelet transform implemented using the filters $\{h_o(n), h_1(n)\}$ on the rows and $\{g_o(n), g_1(n)\}$ on the columns. If the imaginary part of $\Psi(x)\Psi^*(y)$, $\varphi(x)\Psi(y)$, $\Psi(x)\varphi(y)$, $\varphi(x)\Psi^*(y)$, and $\Psi(x)\varphi^*(y)$, is considered, then the six oriented wavelets are given by

$$\Psi_i(x, y) = \frac{1}{\sqrt{2}}(\Psi_{3,i}(x, y) + \Psi_{4,i}(x, y)) \tag{6.15}$$

$$\Psi_{i+3}(x, y) = \frac{1}{\sqrt{2}}(\Psi_{3,i}(x, y) - \Psi_{4,i}(x, y)) \tag{6.16}$$

for $i = 1, 2, 3$, where the two separable 2D wavelet bases are defined as

$$\Psi_{3,1}(x, y) = \varphi_g(x)\Psi_h(y), \quad \Psi_{3,2}(x, y) = \Psi_g(x)\varphi_h(y) \tag{6.17}$$

$$\Psi_{3,3}(x, y) = \Psi_g(x)\Psi_h(y), \quad \Psi_{4,1}(x, y) = \varphi_h(x)\Psi_g(y) \tag{6.18}$$

$$\Psi_{4,2}(x, y) = \Psi_h(x)\varphi_g(y), \quad \Psi_{4,3}(x, y) = \Psi_h(x)\Psi_g(y) \tag{6.19}$$

These six imaginary parts of the complex wavelet sub-bands of the 2D DT-CWT are also strongly oriented in $\{+15°, +45°, +75°, -15°, -45°, -75°\}$ directions.

However a set of six complex wavelets can be formed by using wavelets in Eqs. (6.9)–(6.10) as the real part and wavelets in Eqs. (6.15)–(6.16) as the imaginary parts. The real and imaginary part of each complex wavelet are oriented at the same angle.

Spatial Domain Structural Similarity Algorithm

Natural image signals are highly structured: their pixels exhibit strong dependencies, especially when they are spatially proximate, and these dependencies carry important information about the structure of the objects in the visual scene. The spatial domain structural similarity algorithm (SSIM) has been proposed by Wang et al. [102]. The fundamental principle of the structure approach is that the human visual system is highly adapted to extract structural information (the structure of the objects) from the visual scene. The luminance of the surface of an object being observed is the product of illumination and the reflectance , but the structure of the objects in the scene are independent of the illumination. Wang et al. [102] showed that the structural information in an image are those attributes that represent the structure of objects in the scene, independent of average luminance and contrast.

In the spatial domain, the SSIM index between two image patches $x = \{x_i, i = 1, \ldots, M\}$ and $y = \{y_i, i = 1, \ldots, M\}$, is defined as

$$S(x, y) = \frac{(2\mu_x\mu_y + C_1)(2\sigma_{xy} + C_2)}{(\mu_x^2 + \mu_y^2 + C_1)(\sigma_x^2 + \sigma_y^2 + C_2)} \tag{6.20}$$

where C_1 and C_2 are two small positive constants to avoid instability,

$$\mu_x = \frac{1}{M}\sum_{i=1}^{M} x_i, \quad \sigma_x^2 = \frac{1}{M}\sum_{i=1}^{M}(x_i - \mu_x)^2$$

and

$$\sigma_{xy} = \frac{1}{M}\sum_{i=1}^{M}(x_i - \mu_x)(y_i - \mu_y)$$

respectively.

Here μ_x, σ_x, σ_{xy} are computed within 11×11 circular-symmetric Gaussian window function, then weighted average of all such local windows are obtained to form the SSIM index. It can be seen that the maximum SSIM index value 1 is achieved if and only if x and y are identical.

6.3.2 Extracting Structure Feature of an Image Using Dual Tree Complex Wavelet Transform

The possibility of using dual tree complex wavelet transform to find out structure features invariant to disturbances caused due to illumination and motion in shot boundary detection has been explored. The process of extracting structure feature of an image is described in a flowchart shown in Figure 6.2. The decomposition of an input image with 2D dual tree complex wavelet transform can be obtained by filtering a given image with 2D complex wavelet filters, as described in Eqs. (6.9)–(6.10) to find out the six sub-bands of real parts (denoted by S_R), and Eqs. (6.15)–(6.16) to find out the six sub-bands of imaginary parts (denoted as S_I). The image is decomposed into 12 bandpass oriented sub-bands using 2D DT-CWT for first level of decomposition. These 12 sub-bands gives information strongly oriented at $\{+15°, +45°, +75°, -15°, -45°, -75°\}$ directions for six real and six imaginary sub-bands. Then the magnitude of corresponding real and imaginary

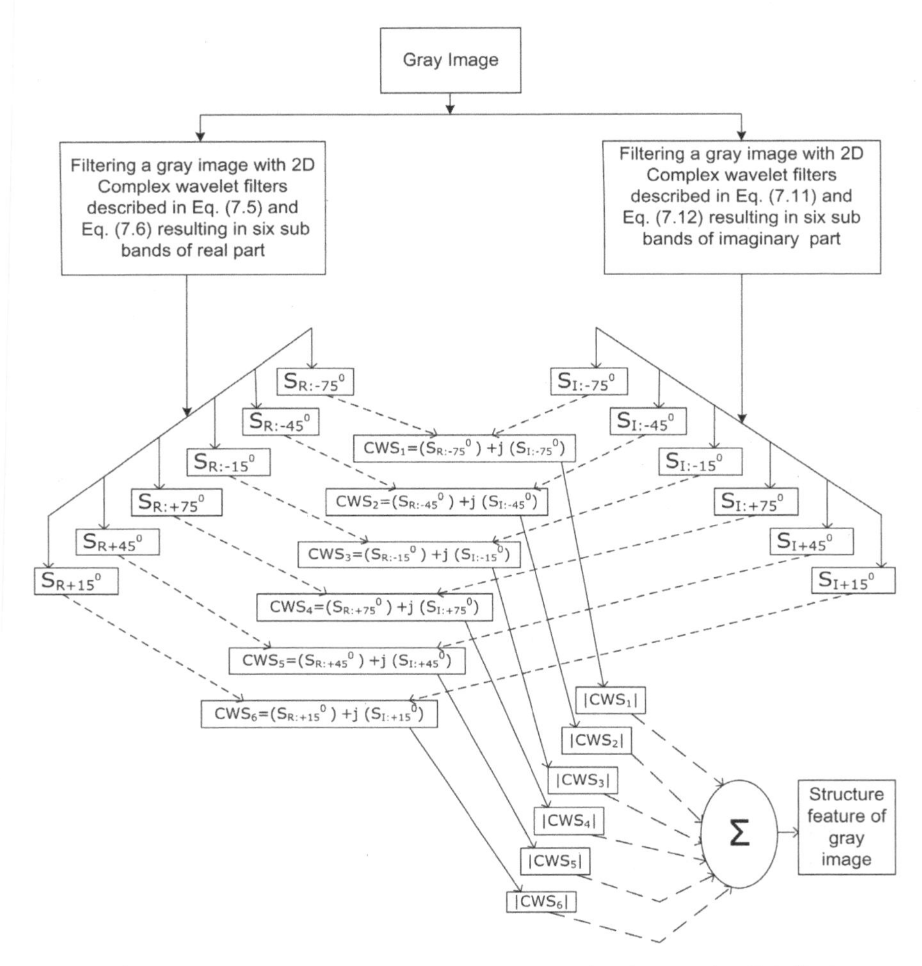

Figure 6.2 Process of extracting structure feature of an image using DT-CWT

coefficients of each sub-bands are obtained (denoted as $|CWS|$). These six magnitude sub-bands of an image are combined to form the structure feature of an image.

To demonstrate the invariance of these features for illumination and motion, four different type of case examples has been considered as described below. Each image is converted from RGB to YUV color space, and only the luminance (Y) component (gray image) is used to find structure features of each image. The structure features of each gray image is obtained as per the process described in Figure 6.2.

Case I: Demonstration of invariance to fire and flicker: Six consecutive frames in the presence of fire and flicker from the movie X-Men (shown in Figure 6.3(a)) have been considered to demonstrate invariance to fire and flicker. Figure 6.3(b) shows the structure features obtained using magnitude of the dual tree complex wavelet transform, which clearly indicates that these feature are invariant to fire and flicker.

Case II – Demonstration of invariance to explosion: Six consecutive frames in the presence of explosion from the movie X-Men (shown in Figure 6.4(a)) have been considered to demonstrate invariance to explosion. Figure 6.4(b) shows the structure features obtained using magnitude of the dual tree complex wavelet transform, which are invariant to explosion.

Case III – Demonstration of invariance to fast object/camera motion: Six consecutive frames in the presence of fast object motion from the movie X-Men (shown in Figure 6.5(a)) have been considered to demonstrate invariance to motion. Figure 6.5(b) shows the structure features obtained using a magnitude of a dual tree complex wavelet transform, which is invariant to fast object motion.

Case IV – Demonstration of invariance to flashlight: In this case we consider six consecutive frames in the presence of flashlight from the movie Sleepy Hollow (shown in Figure 6.6(a)). Figure 6.6(b) shows the structure features obtained using a magnitude of a dual tree complex wavelet transform, which clearly shows that these feature are invariant to flashlight.

For illustrating our proposed method, we considered a small video clip of 117 consecutive frames from the movie X-Men. Here shot boundaries are at frame numbers 3, 14, 29, 44, 51, 60, 89, and 110. In this clip, frames with fast camera and object motion are present in addition to shot boundaries and illumination (due to flashlight, fire, explosion etc.) as shown in Figure 6.7(a). Each frame from the video clip is converted from RGB to YUV color space, and only luminance (Y) component is used to find structure feature of each frame as described in Figure 6.2. These structure features obtained for each frame are used to find potential shot boundaries using the SSIM index algorithm. Let us denote the structure feature obtained for each frame by $\{SF[x, y, 1], SF[x, y, 2], \ldots, SF[x, y, n]\}$, where $\{x = 1, 2, 3, \ldots, M\}$, $\{y = 1, 2, 3, \ldots, N\}$, MN is the frame dimension, and n is the number of frames used from the video for the analysis. For the above example, we use $n = 117$.

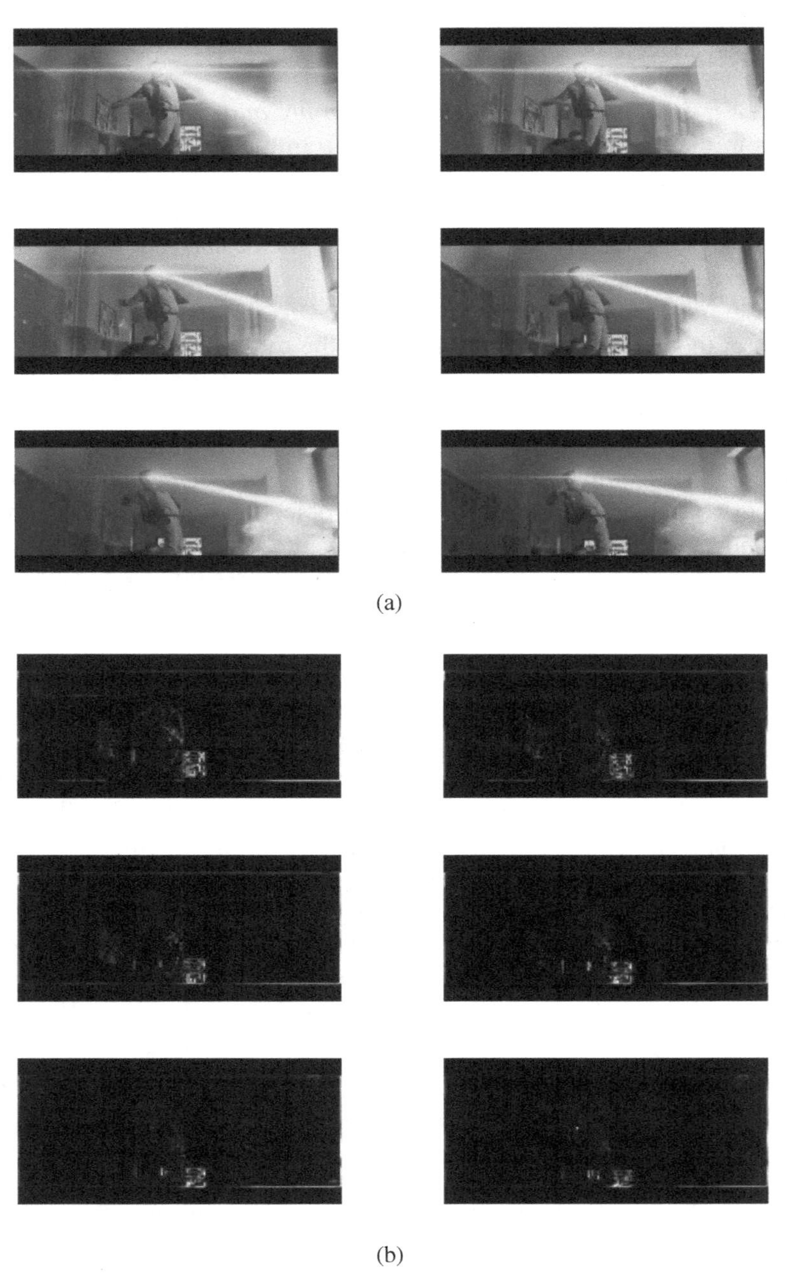

Figure 6.3 (a) Six consecutive frames in the presence of fire and flicker. (b) Structure features obtained using magnitude of the dual tree complex wavelet transform

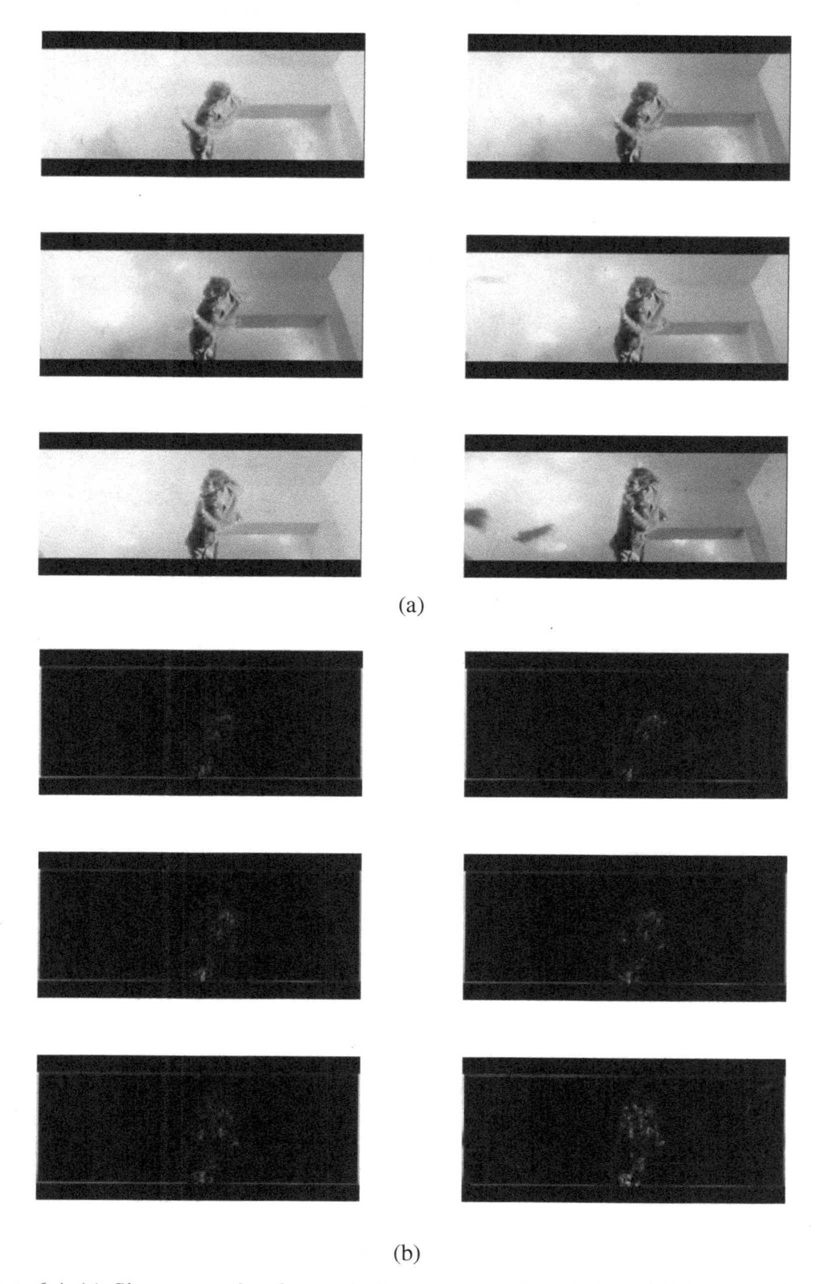

Figure 6.4 (a) Six consecutive frames in the presence of explosion. (b) Structure features obtained using a magnitude of a dual tree complex wavelet transform

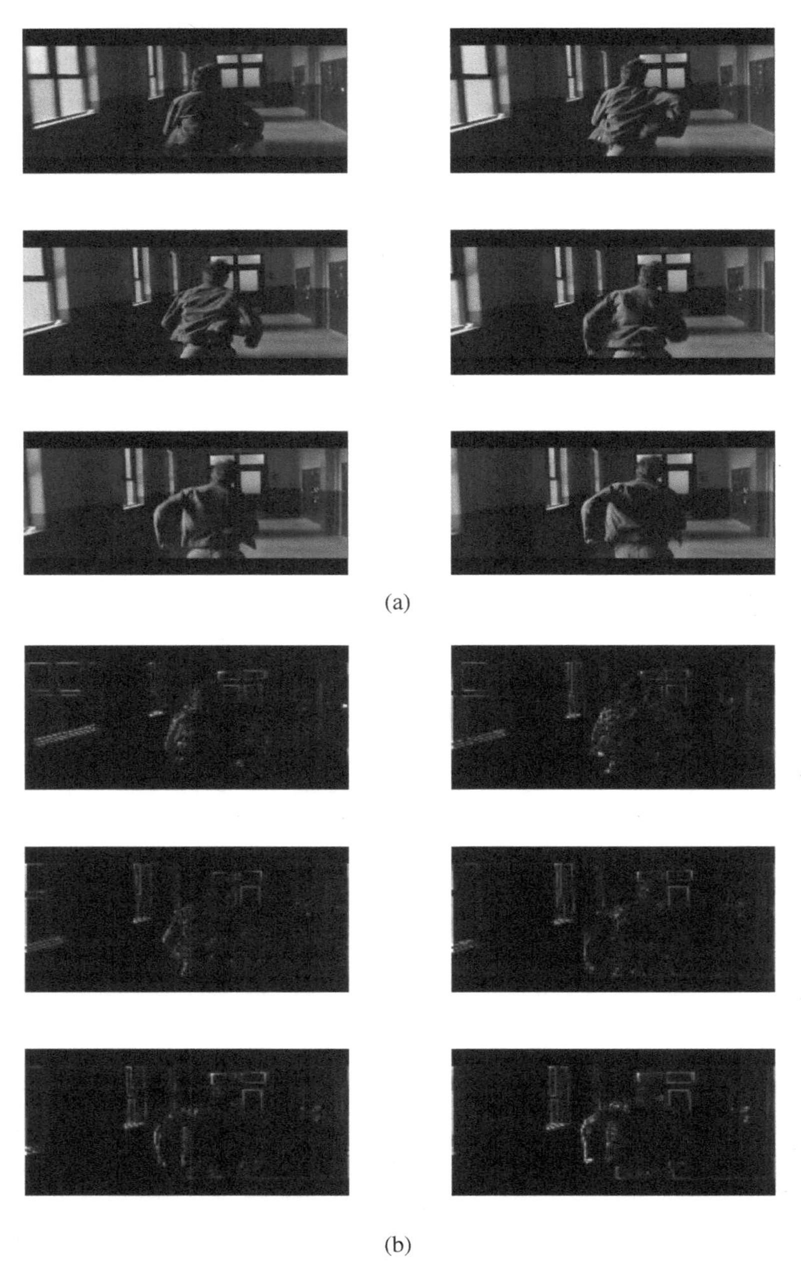

Figure 6.5 (a) Six consecutive frames in the presence of fast object motion. (b) Structure features obtained using a magnitude of a dual tree complex wavelet transform

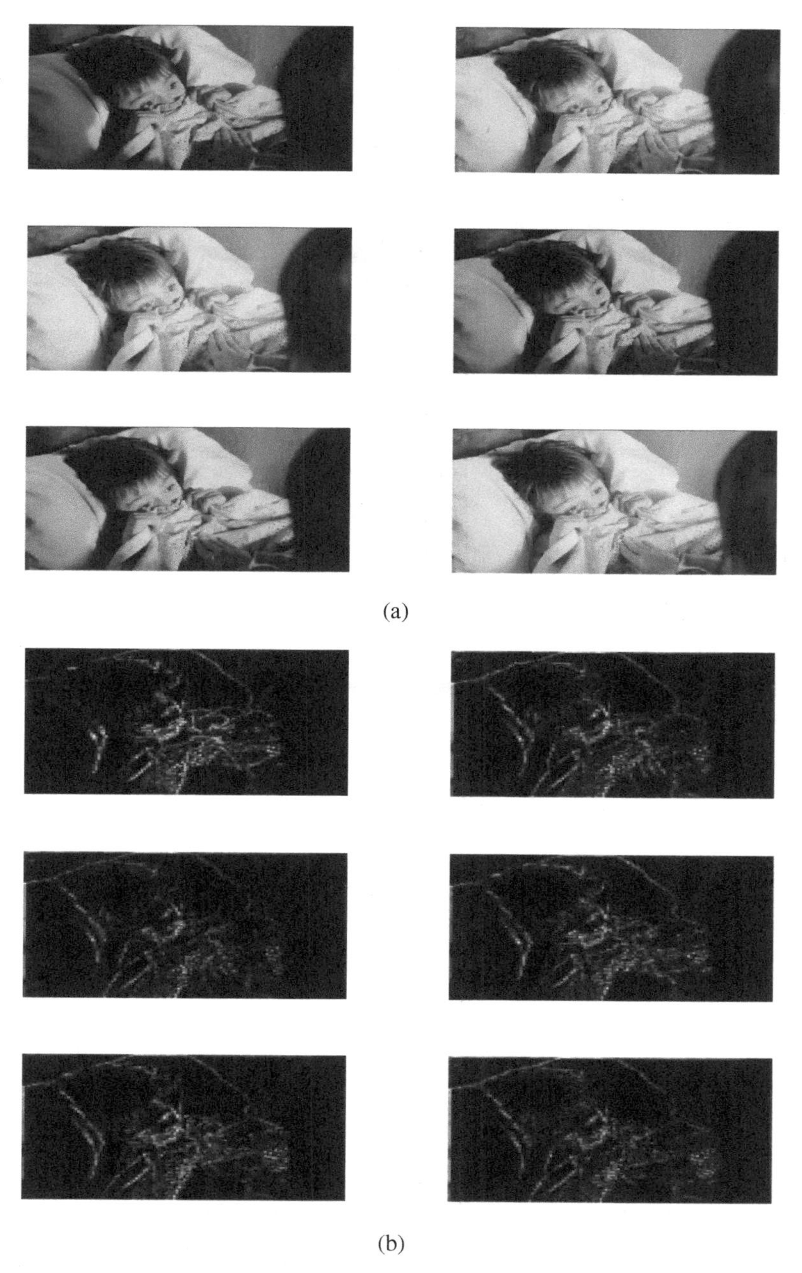

Figure 6.6 (a) Six consecutive frames in the presence of flashlight. (b) Structure features obtained using a magnitude of a dual tree complex wavelet transform

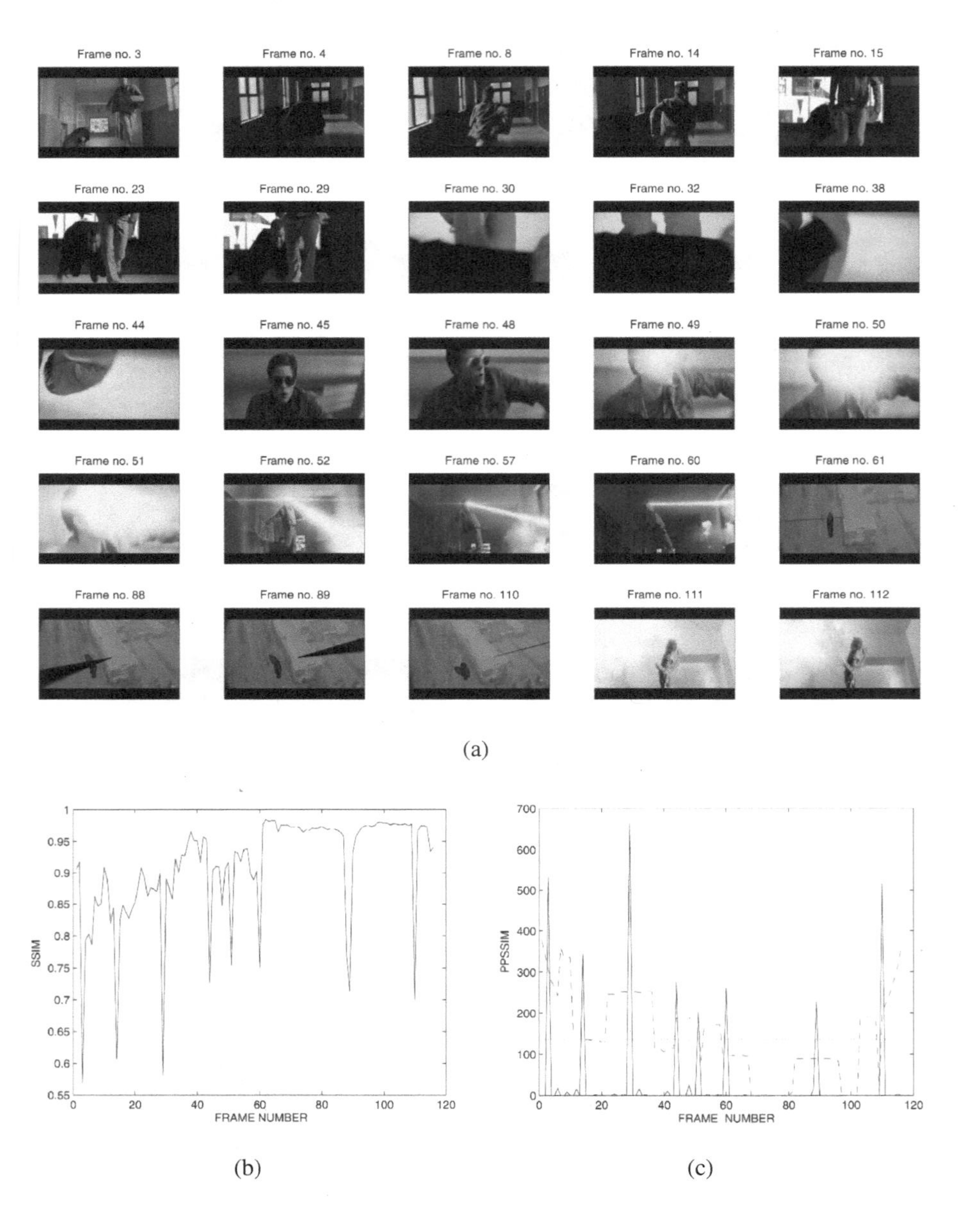

Figure 6.7 (a) Video clip from the movie X-Men. (b) Potential shot boundaries obtained using SSIM index. (c) Shot boundaries with local (dotted lines) and adaptive threshold (dashed lines) (true shot boundaries are at frame numbers 3, 14, 29, 44, 51, 60, 89, 110)

6.3.3 Finding Potential Shot Boundaries Using the SSIM Index Algorithm

The spatial domain structural similarity (SSIM) algorithm [102] used structure information of the object from the scene which is independent of the average luminance and contrast. SSIM algorithm is very successful in finding image similarity [108], so we explored the possibility of this algorithm as a metric to find potential shot boundaries. As we have already obtained the structure feature of an image, so the SSIM index was found between two consecutive structure feature images, instead of using local window as described in [102]. The SSIM index between consecutive 117 frames were obtained by using Eq. (6.21).

$$SSIM(k, k+1) = \frac{(2\mu_k \mu_{k+1} + C_1)(2\sigma_{k,k+1} + C_2)}{(\mu_k^2 + \mu_{k+1}^2 + C_1)(\sigma_k^2 + \sigma_{k+1}^2 + C_2)} \tag{6.21}$$

for $1 \leq k \leq n-1$, where μ_k is the mean of the structure feature of frame $SF[x, y, k]$ and is obtained by

$$\mu_k = \frac{1}{MN} \sum_{x=1}^{M} \sum_{y=1}^{N} SF[x, y, k]$$

σ_k^2 is obtained by

$$\sigma_k^2 = \frac{1}{MN} \sum_{x=1}^{M} \sum_{y=1}^{N} (SF[x, y, k] - \mu_k)^2$$

$\sigma_{k,k+1}$ is obtained by

$$\sigma_{k,k+1} = \frac{1}{MN} \sum_{x=1}^{M} \sum_{y=1}^{N} (SF[x, y, k] - \mu_k)(SF[x, y, k+1] - \mu_{k+1})$$

C_1 and C_2 are small constants to avoid instability. Any small value of C1 and C2 (less than 1) does not affect the SSIM results, whereas higher values (greater than 1) affect the results of the SSIM index. Hence C_1 and C_2 should be always less than one, preferably very small value. For the simulation we use $C_1 = 0.001$, $C_2 = 0.002$. Figure 6.7(b) shows the SSIM index obtained for 117 consecutive structure feature frames described in Section 6.3.2.

The maximum SSIM index value 1 is achieved when the frames are identical, whereas lower values indicates dissimilarity.

6.3.4 Declaration of Shot Boundaries Using Post Processing, Local Threshold and Adaptive Threshold

The SSIM index coefficients are post processed by using following steps and denoted as PPSSIM:

$$\{PPSSIM(i) = \gamma \times |1 - SSIM(i)|\}, \text{ for } 1 \leq i \leq n-1$$

where γ is a scaling coefficient.

Then the combination of local and adaptive threshold (as discussed in Section 5.3.3) has been applied for shot boundary detection.

The proposed algorithm is applied to a initial video clip of 200 frames and then subsequently applied to the next video clip of 200 frames till all the frames in the test video sequence of one movie are tested. The threshold for initial 200 frames was found out using local threshold (denoted as L_T)) and defined as

$$L_T = \alpha \times \frac{1}{n-1} \sum_{i=1}^{n-1} PPSSIM(i) \tag{6.22}$$

where α is a scaling coefficient.

As L_T depends on the mean value of 200 frames, it will change for the next 200 consecutive frames so this threshold is termed as a local threshold.

As the local threshold is constant for 200 frames and might result in detecting false positives, an adaptive threshold has been defined which will adjust the threshold for each frame using the average PPSSIM value of 6 frames preceding the current frame and the average PPSSIM value of 6 frames after the current frame. This is denoted by A_T and defined by

$$A_T(i) = \beta \times AI(i) \tag{6.23}$$

$$AI(i) = \begin{cases} \frac{1}{6}\sum_{k=i+1}^{i+6} PPSSIM(k) & \text{for } 1 \leq i \leq 6 \\ \frac{1}{2}\{\frac{1}{6}\sum_{k=i-6}^{i-1} PPPSSIM(k) & \\ \quad + \frac{1}{6}\sum_{k=i+1}^{i+6} PPSSIM(k)\} & \text{for } 7 \leq i \leq n-7 \\ \frac{1}{6}\sum_{k=i-6}^{i-1} PPSSIM(k) & \text{for } n-6 \leq i \leq n-1 \end{cases} \tag{6.24}$$

where β is a scaling coefficient, and AI is the average between PPSSIM coefficients in left sliding window preceding the PPSSIM value of a current frame, and PPSSIM coefficients in right sliding window after the PPSSIM value of a current frame as defined in Eq. (6.24).

AI is calculated by the following method. First we put the average PPSSIM value of 6 frames preceding the current frame into the left sliding window. Simultaneously the average PPSSIM value of 6 frames after the current frame is put into the right sliding window. Then we use the average PPSSIM of all frames within the right and left sliding windows to calculate AI. For the first 6 frames, a left sliding window did not exist, hence the average PPSSIM value of a right sliding window was considered. While for the last 6 frames, a right sliding window did not exist, hence the average PPSSIM value of a left sliding window was considered. The results of PPSSIM coefficients with local and adaptive thresholds is shown in Figure 6.7(c). Here the local threshold was denoted by dotted lines and the adaptive threshold by dashed lines. If the PPSSIM values at frame number k is above both the local and adaptive thresholds, then shot boundary is declared at frame number k.

From Figure 6.7(c) it can be clearly observed that the shot boundaries are at frame numbers 3, 14, 29, 44, 51, 60, 89, and 110. Hence the proposed algorithm found shot boundaries correctly in the presence of illumination and motion without any false positive due to these effects.

6.3.5 Demonstration of Results Using the Proposed Algorithm When Fast Camera and Object Motion Is Present

To demonstrate results in the presence of motion, 119 frames from the movie X-Men have been considered from the test video sequence (shown in Figure 6.8(a)). The actual shot boundaries are at frame numbers 13, 25, 51, 73, 93 and 105. In this clip fast camera and object motion is observed in almost all the frames with sufficient number of shot boundaries. We used this clip to show the robustness of our proposed algorithm in the presence of motion.

The potential shot boundaries obtained after applying dual tree complex wavelet transform and SSIM algorithm to this video clip are shown in Figure 6.8(b). The results obtained after applying post processing and thresholds (denoted as PPSSIM) to the SSIM index are shown in Figure 6.8(c), which clearly indicate that all the shot boundaries are correctly detected though fast camera and object motion is present in the consecutive frames.

6.4 Performance Comparison of the Proposed Algorithm with Other Shot Boundary Detection Methods

6.4.1 Test Video Sequence and an Evaluation Criterion

The proposed algorithm has been tested on two movies X-Men (denoted as XM) and Home Alone (denoted as HA). These movies are manually observed

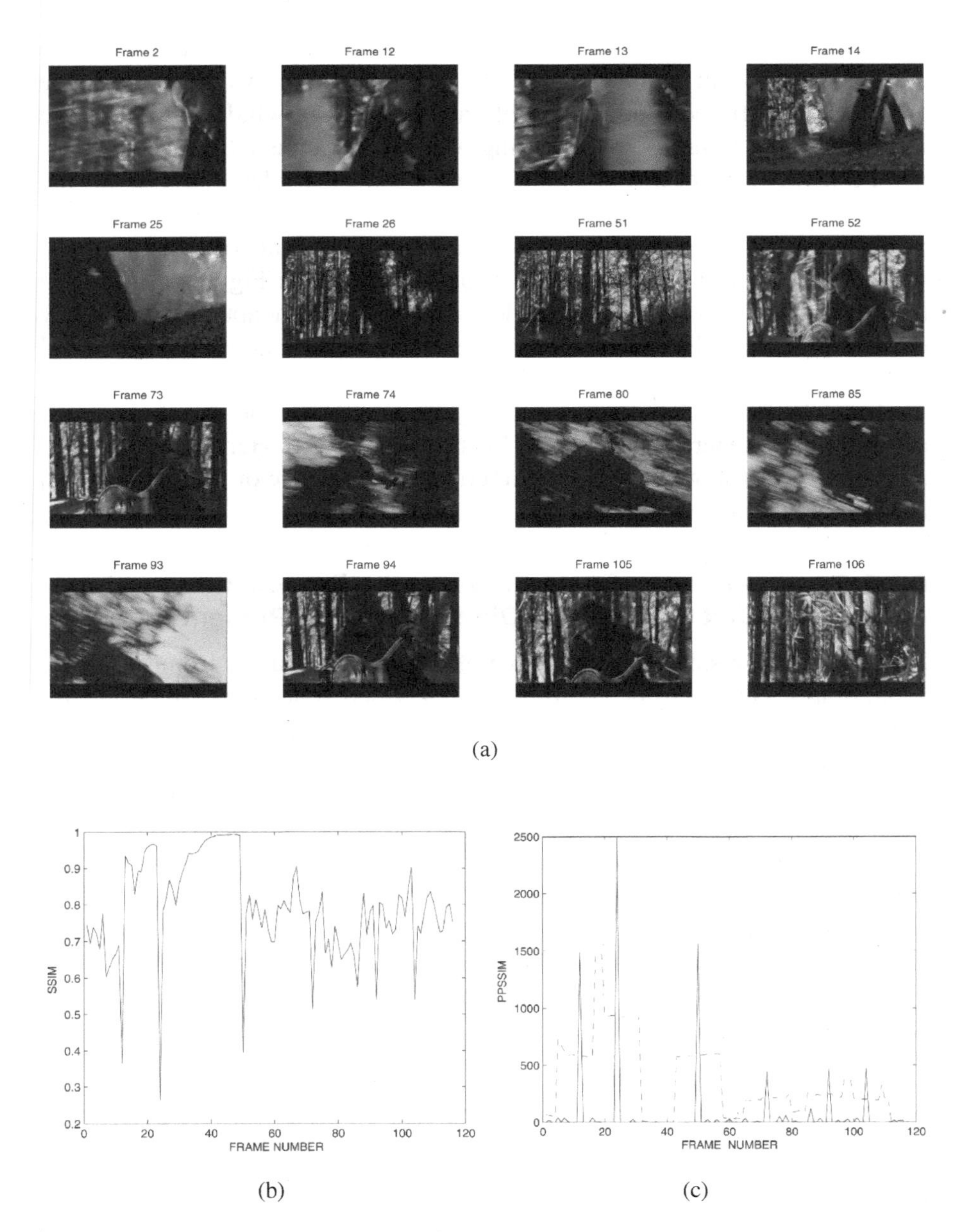

Figure 6.8 (a) Video clip from the movie X-Men. (b) Potential shot boundaries obtained using SSIM index. (c) Shot boundaries with local (dotted lines) and adaptive threshold (dashed lines) (true shot boundaries are at frame numbers 13, 25, 51, 73, 93 and 105)

Table 6.5 Number of shot boundaries, and number of frames with object and camera motion effects used in each movie for the analysis

Movie	XM	HA
Shot boundaries	551	182
Frames with motion	11373	7333

frame by frame to find actual shot boundaries. These movies are considered for obtaining the test data, since large number of frames with object motion and camera motion are present in addition to shot boundaries. The number of shot boundaries and number of frames with object and camera motion effects for each movie is shown in Table 6.5. The test data also contain frames with flashlight and FFE.

We used Recall, Precision and F1 measure as discussed in Section 2.5 as an evaluation metric to compare shot boundary detection algorithms.

6.4.2 Performance Comparison

The performance of the proposed algorithm was compared with the methods developed by Cheol et al. [2] and Yang et al. [35] for the same video sequence. These algorithms were tested on two movies X-Men and Home Alone. The frame based R, P, and F1 measure performance comparison between the proposed and the other tested algorithm are shown in Tables 6.6 and 6.7 for the movie X-Men and Home Alone respectively. The scaling parameter γ, α (Eq. 6.22), and β (Eq. 6.23) were empirically chosen.

Cheol et al. [2] proposed an algorithm for robust change detection in the presence of flashlight. In this method local color histogram, robust to small object and camera motion was used to find the difference between consecutive frames. This difference is dynamically compressed by logarithmic transform and then four threshold conditions are used to detect shot boundary. This algorithm is robust to small object motion and weak flashlight, but produces false positives for camera motion, fast object motion, and rapidly varying illumination.

The motion suppression technique for shot boundary detection based on 3D wavelet transform have been proposed by Yang et al. [35]. This algorithm was chosen as it suppress motion with respect to shot boundaries, and also robust to flashlight, so we used it for comparison with the proposed method. They used higher frequency spatio-temporal edge information after 3D wavelet decomposition so as to make metric robust to illumination changes. Then intensity of motion energy is extracted and used to suppress motion influence.

Table 6.6 Performance comparison of the proposed algorithm for the movie X-Men

Algorithm	D	C	M	FP	R	P	F1
Cheol et al. [2]	551	468	83	188	84.93	71.31	77.52
Yang et al. [35]	551	461	90	163	83.66	73.87	78.46
Proposed algorithm	551	522	29	27	94.73	95.08	94.90

Table 6.7 Performance comparison of the proposed algorithm for the movie Home Alone

Algorithm	D	C	M	FP	R	P	F1
Cheol et al. [2]	182	168	14	56	92.31	75	82.75
Yang et al. [35]	182	167	15	47	91.76	78.08	84.34
Proposed algorithm	182	180	02	03	98.90	98.36	98.62

This algorithm is able to suppressed the slow motion with respect to shot boundaries, but unable to do so for the motion caused by fast camera and object. This algorithm also produced false positives for nonuniform flashlight condition.

The efficacy of the proposed method can be clearly observed in Tables 6.6 and 6.7; it outperformed other methods in terms of better Recall (R) and Precision (P). Proposed algorithm is also robust in the sense that, our test data contains large number of frames with fast camera and object motion compared with abrupt transition as shown in Table 6.5. Also a considerable number of frames with flashlight and FFE is present in the test video sequence. Other methods used for comparison are unable to avoid false positives caused due to fast motion on the same data, whereas our method provides better trade-off between Recall and Precision. However, the performance of the proposed method is sensitive to fast camera motion, especially when the object is in/out during the motion in the same scene or during the rapid change in background, whereas in the case of the camera following a fast moving object, this algorithm is able to avoid false positives due to these motions.

6.5 Summary and Conclusions

Disturbances caused by illumination change or fast object and camera motion are often mistaken as shot boundaries and its elimination are the major challenge to the shot boundary detection algorithms. We evaluated the performance of major traditional shot boundary algorithms in the presence of motion for various color space. From the experimental results, it has been found that the color histogram metric performed better than the likelihood ratio in RGB color space, whereas likelihood ratio performed better than histogram difference in HSV color space. While in YUV color space, the chi-square method performed better than pixel difference metric.

The performance of all the metric is poor due to the disturbances caused by fast camera and object motion. Hence, we propose an algorithm for shot boundary detection in the presence of motion using dual tree complex wavelet transform. The poor performance of the traditional shot boundary detection metrics caused by disturbance of motion is overcome by our proposed algorithm.

In this chapter, we addressed two major issues by using the structure features which are invariant to illumination and motion using dual tree complex wavelet transform. A spatial domain structure similarity algorithm, which is also invariant to average luminance and contrast has been applied to these structure features to detect shot boundaries. Finally local and adaptive thresholds have been used to declare correct shot boundaries.

Our proposed algorithm has been tested on action movie data, where large number of frames with fast object and camera motion are observed in addition to frames with flashlight, fire, flicker and explosion. Experimental results are carefully evaluated using the performance metric Recall, Precision, and F1 measure. Our proposed algorithm achieved a relatively better trade-off between Recall and Precision with high F1 measure as compared to other tested algorithms on the same video data. The proposed algorithm is successful in avoiding disturbances due to illumination change and fast motion when camera follow the object. However our method is sensitive to unusual cases when background in consecutive frames changes rapidly in addition to appearance and disappearance of multiple objects in the same scene.

7

Conclusion and Future Work

7.1 Conclusion on the Proposed algorithms for Shot Boundary Detection

In this monograph, we have addressed some issues in the shot boundary detection. We summarize our results below.

7.1.1 Effective Algorithm for Detecting Various Wipe Patterns

We addressed the problem of wipe transition detection by proposing an algorithm for detecting various wipe patterns. In the proposed algorithm, the moving strip due to wipe was obtained, followed by an application of the Hough transform to these moving lines so as to detect and categorize various wipe types. Applying the proposed algorithm to the whole video to detect wipes is computationally expensive. In order to decrease the computational load of the proposed algorithm, a preprocessing step as the first stage of the algorithm has been proposed. The proposed algorithm with the preprocessing step depends on the tuning parameters, requiring less computation with a slight degradation in F1 score, while without a preprocessing step, it would be a general method independent of the tuning parameters involving more computations.

Our proposed algorithm has been extensively tested on several genres of video data (with and without preprocessing). Experimental results were evaluated using the performance metrics Recall, Precision, F1 measure, and Detection Rate. Our proposed algorithm achieved a relatively better trade off between Recall and Precision as compared to other algorithms. The selection of the tuning parameters and its effect on performance of the results in terms of Recall, Precision and F1 score in the preprocessing stage has also been discussed. We have tested the proposed approach on 31 different wipe effects including 18 complex and special wipes, and obtained Detection Rate of 94.12, 100, 100, 100, and 96.67 for Star War III, Star War I, Jodhaa

Akbar, Bhooth Nath and Tomcat movies, respectively. The proposed method successfully avoided false positives caused due to object and camera motion, by differentiating wipe transitions and motion using various gradient patterns. The proposed algorithm was able to detect 28 wipe types out of 31 wipe types, found in this test video sequence and resulted in high Recall and higher Detection Rate. Our algorithm failed to detect wipe types VWP3, VWP4 and Tom8 due to its nonlinear and nonrigid changing patterns and boundary. The Hough transform failed to find differentiable gradient pattern for these wipe types. Our algorithm was able to find out the wipe range accurately in most of the wipe types excluding HWP7 and diagonal wipes, where scene change region is very small during starting and end of the wipe. This is the main reason for high precision in the proposed algorithm. The proposed algorithm also discriminates wipes from object and camera motion. Overall our proposed algorithm gave better tradeoff between Recall and Precision as compared to other compared algorithms and detected most of the wipe types and their range accurately. Our proposed algorithm successfully differentiated between horizontal, vertical, diagonal and special wipe types by observing gradient patterns satisfied by them.

7.1.2 Shot Boundary Detection in the Presence of Flashlights

An effective method has been proposed for shot boundary detection in the presence of flashlights. The proposed algorithm suppressed the effect due to flashlight by using logarithmic transform followed by discrete cosine transform. Discrete wavelet transform based metric in combination with local and automatic threshold was used to find shot boundaries. The performance of the proposed algorithm was tested on the movies Sleepy Hollow, Independence Day and Deep Rising and results were compared with other shot boundary detection methods in the presence of flashlight. The selection of the tuning parameters and its effect on performance of the results in terms of Recall, Precision and F1 score was also discussed.

The proposed algorithm is robust in the sense that our test data consist of 10–30 times more frames with flashlight than frames with abrupt transition and also have a combination of weak and strong flashlights. Other methods used for comparison are unable to avoid false positives caused due to flashlight on the same data, whereas our method provides better trade-off between Recall and Precision. However, the performance of the proposed method is sensitive to large camera/object motion and results in false positives. Main goal of this work is to suppress the disturbances caused by flashlight which is one of the major challenge in shot boundary detection and we achieve it reasonably well. Our test data contains mostly flashlight with camera breaks and reasonably small object/camera motion (observed in the Sleepy Hollow movie) and few frames with large camera/object motion (observed in Independence movie). The Illumination Suppression algorithm using logarithmic transform and DCT has resulted in suppression of flashlights, whereas wavelet transform based metrics in combination with adaptive threshold have been successful in avoiding the effect of small camera and object motion. The first stage of our proposed method which is used to suppress flashlight effect using logarithmic transform and discrete wavelet transform can be a preprocessing step for other metrics and may reduce the false positives due to flashlights.

7.1.3 Shot Boundary Detection in the Presence of Fire Flicker and Explosion

Detection of shot boundaries in the thriller movies under FFE effects is a difficult task. The major problem arises from the fact that changes in luminance is not uniform over consecutive frames during this effect, as the direction and position of fire moves in the consecutive frames. In few cases, appearance and disappearance of fire in the consecutive frames contributes to the large differences. The effectiveness of various metrics under FFE effects have been evaluated and found that, color ratio histogram and cross correlation coefficient metric performed better than the other compared metrics under these effects. These traditional shot boundary detection algorithms are sensitive to disturbances caused by camera motion, object motion and sudden illumination changes, and produced false positives for such cases. The behavior of FFE under which these metrics failed has been discussed. We have proposed an effective algorithm based on cross correlation coefficient, stationary wavelet transform and combination of local and adaptive thresholds for shot boundary detection. Experimental results on thriller movies Pearl Harbor, The Marine and Saving Private Ryan show significant improvements in terms of better Recall and Precision. The proposed algorithm is found to achieve a high F1 measure compared to other tested algorithms. The false positives detected by our proposed algorithm were due to the disturbances caused by fast camera and object motion.

7.1.4 Performance Evaluation of Traditional Metrics in the Presence of Motion

We evaluated the performance of major traditional shot boundary algorithms in the presence of motion in various color space. From the experimental results it has been found that color histogram perform better than likelihood ratio in RGB color space, whereas likelihood ratio performs better than histogram difference in HSV color space. In YUV color space chi-square method perform better than pixel difference metric. Overall it has been observed that all these metrics have provided poor results due to disturbances caused by fast camera and object motion. The maximum false positives and missed detections are due to frame difference between consecutive frames caused by fast camera motion.

7.1.5 Shot Boundary Detection in the Presence of Illumination and Motion

Disturbances caused by illumination change or fast object and camera motion are often mistaken as a shot boundaries and its elimination is the major challenge to the shot boundary detection algorithms. We addressed these two issues by developing the structure features (using dual tree complex wavelet transform) which are invariant to illumination and motion. Then, the spatial domain structure similarity algorithm, which is also invariant to average luminance and contrast, has been applied to these structure features to detect shot boundaries. Finally, a local and adaptive threshold is used to declare correct shot boundaries.

Our proposed algorithm has been tested on action movie X-Men and Home Alone, where large number of frames with fast object and camera motion are observed in addition to frames with flashlight, fire, flicker and explosion. Experimental results were carefully evaluated using the performance metric Recall, Precision, and F1 measure. Our proposed algorithm achieved a relatively better trade-off between Recall and Precision with high F1 measure compared to other tested algorithms on the same video data. Other methods used for comparison are unable to avoid false positives caused due to fast motion on the same data, whereas our method provided better trade-off between Recall and Precision in such cases. The proposed algorithm is successful in avoiding disturbances due to illumination change and fast motion when the camera follows the object. However our method is sensitive to unusual cases when background in consecutive frames changes rapidly in addition to appearance and disappearance of multiple objects in the same scene.

7.2 Directions for Further Research

In this book, we proposed algorithms for wipe detection and eliminating the disturbances due to flashlight, fire, flicker, explosion and motion. However, some issues that need to be addressed in the future are:

- It is a challenging task to develop an algorithm, which is not only invariant to various disturbances such as illumination and motion but also having excellent detection performance for all types of shot boundaries. Hence it could be seen as a future task to develop a single metric which will solve these issues.
- In order to achieve high accuracy in shot boundary detection, an appropriate threshold must be chosen. As the video contents could change dramatically in the consecutive frames of the movie, it is challenging task to find a threshold that works with all kinds of video material. Probabilistic detection or trained classifier such as SVM, HMM and k-means clustering can be used for deciding shot boundaries in future SBD algorithms.
- Differentiating and categorizing detected shot boundaries as abrupt transition, dissolve, wipe, fade-in, fade-out and digital video effects is also a challenging task and all the possibilities in this context could not be explored. A thorough investigation into such possibilities needs to be carried out.
- Issues such as real time implementation and computational complexities in SBD have received less attention from the research community.
- An interesting area that would involve further research would be gesture based or object based scene change detection and may be very useful for gesture or object based video retrieval.

Bibliography

[1] Zhang D., Qi W., and Zhang H.J., A new shot boundary detection algorithm. In *PCM*, Lecture Notes in Computer Science, Vol. 2195, pp. 63–70. Springer-Verlag, 2001.

[2] Cheol K., Cheon Y., Kim G., and Choi H., Robust scene change detection algorithm for flashlights. In *ICCSA*, Lecture Notes in Computer Science, Vol. 4705, pp. 1003–1013. Springer-Verlag, 2007.

[3] Zhang H.J., Kankanhalli A., and Smoliar S., Automatic partitioning of full-motion video. *Multimedia Systems*, 1(1):176–189, 1993.

[4] Li D. and Lu H., Avoiding false alarms due to illumination variation in shot detection. In *Proceedings of IEEE Workshop on Signal Processing Systems*, pp. 828–836, 2000.

[5] Nagasaka A. and Tanka Y., Automatic video indexing and full video search for object appearance. In *Visual Database Systems II*, E. Knuth and L. Wegner (Eds.), pp. 113–127. Elsevier Science Publishers, 1992.

[6] Boreezky J.S. and Rowe L.A., Comparison of video shot boundary detection techniques. In *Proceedings of SPIE Conference on Storage and Retrieval for Image and Video Databases*, Vol. 2670, No. IV, pp. 170–179, 1996.

[7] Ahanger G. and Little D.C., A survey of technologies for parsing and indexing digital video. *Journal of Visual Communication and Image Representation*, 7(1):28–43, 1996.

[8] Lienhart R., Comparison of automatic shot boundary detection algorithms. *SPIE Image and Video Processing*, No. VII, pp. 25–30, 1999.

[9] Gargi U., Kasturi R., and Strayer S., Performance characterization of video-shot-chnage detection methods. *IEEE Transaction on Circuits and Systems for Video Technology*, 10(1):1–13, 2000.

[10] Hanjalic A., Shot-boundary detection unraveled and resolved. *IEEE Transaction on Circuits and System for Video Technology*, 12(2):90–105, 2002.

[11] Yuan J., Wang H., and Xiao L., A formal study of shot boundary detection. *IEEE Transactions on Circuits and System for Video Technology*, 17(2):168–186, 2007.

[12] Smeaton A.F., Over P., and Doherty A.R., Video shot boundary detection: Seven years of TRECVid activity. *Journal of Computer Vision and Image Understanding*, 114(4):411–418, 2010.

[13] Alattar A.M., Wipe scene change detector for use with video compression algorithm and MPEG-7. *IEEE Transactions on Consum. Electron.*, 44(1):43–51, 1998.

[14] Fernando W.A.C., Canagarajah C.N., and Bull D.R., Wipe scene change detection in video sequences. In *Proc. ICASSP*, pp. 294–298, 1999.

[15] Drew M.S., Li Z-N, and Zhang X., Video dissolve and wipe detection via spatio-temporal images of chromatic histogram differences. In *Proceedings of IEEE International Conference on Image Processing*, Vol. 3, pp. 929–932, 2000.

[16] Nam J. and Tewfik A.H., Detection of gradual transitions in video sequences using B-Splines interpolation. *IEEE Transactions Multimedia*, 7(4):667–679, 2005.
[17] Iwamoto K. and Hirata K., Detection of wipes and digital video effects based on a pattern-independent model of image boundary line characteristics. *ICIP*, 6:297–300, 2007.
[18] Li S. and Lee M., Effective detection of various wipe transitions. *IEEE Transactions on Circuits and Systems for Video Technology*, 17(6):663–673, 2007.
[19] Alattar A.M., Detecting fade regions in uncompressed video sequences. In *Proceedings of IEEE International Conference on Acoustics, Speech, and Signal Processing*, pp. 3025–3028, 1997.
[20] Lienhart R., Reliable dissolve detection. In *Proceedings of SPIE Storage and Retrieval for Media Databases*, Vol. 4315, pp. 219–230, 2001.
[21] Albanese M., Chianese A., Moscato V., and Sansone L., A formal model for video shot segmentation and its application via animate vision. *Multimedia Tools and Applications*, 24:253–272, 2004.
[22] Rong J., Ma Y-F, and Wu L., Gradual transition detetcion using EM curve fitting. In *Proceedings of the Eleventh International Multimedia Modelling Conference, MMM'05*, 2005.
[23] černeková Z. and Pitas I., Information theory-based shot cut/fade detection and video summarization. *IEEE Transaction on Circuits and Systems for Video Technology*, 16(1):82–91, 2006.
[24] Weixin K., Ding X., Lu H., and Songde M, Improvement of shot detection using illumination invariant metric and dynamic threshold selection. In Lecture Notes in Computer Science, Vol. 2195, pp. 63–70. Springer-Verlag, 2001.
[25] Guimaraes S., Couprie M., Araujo A., and Leite N., Video segmentation based on 2D image analysis. *Pattern Recognition Letters*, 24(7):947–957, 2003.
[26] Heng W.J. and Ngan K.N., High accuracy flashlight scene determination for shot boundary detection. *Signal Processing: Image Communication*, 18(3):203–219, 2003.
[27] Marbach G., Loepfe M., and Brupbacher T., An image processing technique for fire detection in video images. *Fire Safety Journal*, 41(4):285–289, 2006.
[28] Ugur Treyin B., Yigithan Dedeoglu, Ugur Güdükbay, and Enis **c**Cetin A., Computer vision based method for real time fire and flame detection. *Pattern Recognition Letters*, 27(1):49–58, 2006.
[29] Celik T., Demirel H., and Ozkaramanli H., Fire detection using statistical color model in video sequences. *Journal of Visual Communication and Image Representation*, 18(2):176–185, 2007.
[30] Vinicius P., Borges K., Mayer J., and Izquierdo E., Efficient visual fire detection applied for video retrieval. In *Proceedings of Sixteenth European Signal Processing Conference (EUSIPCO)*, Lausanne, Switzerland, pp. 25–29, 2008.
[31] Celik T. and Demirel H., Fire detection in video sequences using a generic color model. *Fire Safety Journal*, 44(2):147–158, 2009.
[32] Lienhart R., Reliable transition detection in videos: A survey and practitioner's guide. *Int. J. Image Graph*, 1(3):469–486, 2001.
[33] Lawrence S., Ziou D., and Wang S., Motion insensitive detection of cut and gradual transitions in digital video. In *Proceedings of International Conference on Multimedia Modelling*, Ottawa, 1999.

[34] Su C., Liao H., Fan K., and chen L., A motion-tolerant dissolve detection algorithm. *IEEE Transaction on Multimedia*, 7(6):1106–1113, 2005.

[35] Xu Y., De X., Tengfei G., Aimin W., and Congyan L., 3-DWT based motion suppression for video shot boundary detection. In *Proceedings KES 2005*, R. Khosla et al. (Eds.), Lecture Notes in Artificial Intelligence, Vol. 3682, pp. 1204–1209. Springer-Verlag, 2005.

[36] Jang S., Kim G., and Choi H., Shot transition detection by compensating for global and local motions. In *Proceedings FSKD 2005*, L. Wary and Y. Jin (Eds.), Lecture Notes in Arfiticial Intelligence, Vol. 3614, pp. 1061–1066. Springer-Verlag, 2005.

[37] Park M., Park R., and Lee S., Efficient shot boundary detection for action movies using blockwise motion-based features. In *Proceedings ISVS 2005*, G. Bebis et al. (Eds.), Lecture Notes in Computer Science, Vol. 3804, pp. 478–485. Springer-Verlag, 2005.

[38] Jain R., Kasturi R., and Schunck B., *Machine Vision*. McGraw-Hill, New York, 1995.

[39] Sethi I. K. and Patel N., A statistical approach to scene change detection. In *SPIE Proceedings on Storage and Retrieval for Image and Video Databases*, Vol. 2420, pp. 329–338, 1995.

[40] Albiol A., Naranjo V., and Angulo J., Low complexity cut detection in the presence of flicker. *Proc. International Conference on Image Processing*, Vol. 3, pp. 957–960, 2000.

[41] Hsu P.R. and Harashima H., Detecting scene changes and activities in video databases. In *Proceedings of ICASSP'94*, Vol. 5, pp. 33–36, 1994.

[42] Shahraray B., Scene change detection and content-based sampling of video sequences. In *Proceedings of SPIE Conference on Digital Video Compression: Algorithms and Technologies*, Vol. 2419, pp. 2–13, 1995.

[43] Hamapur A., Jain R., and Weymouth T., Production model based digital video segmentation. *Multimedia Tools and Applications*, 1(1):9–46, 1995.

[44] Zabih R., Miller J., and Mai K., A feature-based algorithm for detecting and classifying scene breaks. In *Proceedings of ACM Multimedia*, San Francisco, CA, pp. 189–200, 1995.

[45] Yeo B. and Liu B., Rapid scene analysis on compressed video. *IEEE Transaction on Circuits and Systems for Video Technology*, 5(6):533–544, 1995.

[46] Idris F. and Panchanathan S., Review of image and video indexing techniques. *Journal of Visual Communication and Image Representation*, 8(2):146–166, 1997.

[47] Ford R., Roboson C., Temple D., and Gerlach M., Metrics for shot boundary detection in digital video sequences. *Multimedia System*, 8:37–46, 2000.

[48] Becós J., Cisneros G., Martínez J., and Cabrera J., A unified model for techniques on video-shot transition detection. *IEEE Transaction on Multimedia*, 7(2):293–307, 2005.

[49] Cotsaces C., Nikolaidis N., and Pitas I., Video shot boundary detection and condensed representation: A review. *IEEE Signal Processing Magazine*, 23(2):28–37, 2006.

[50] Alan F. Smeaton, Techniques used and open challenges to the analysis indexing and retrieval of digital video. *Pattern Recognition Letters*, 32, 2007.

[51] Wu M., Wolf W., and Liu B., An algorithm for wipe detection. In *Proceedings of ICIP*, pp. 893–897, 1998.

[52] Kim H., Lee J., and Song S., An efficient graphical shot verifier incorporating visual rhythm. *Proceedings of IEEE International Conference on Multimedia Computing and Systems*, Vol. 1, pp. 827–834, 1999.

[53] Pei S.C. and Chou Y.Z., Effective wipe detection in MPEG compressed video using macroblock type information. *IEEE Transaction on Multimedia*, 4(3):309–319, 2002.

[54] Campisi P., Neri A., and Sorgi L., Wipe effect detection for video sequences. In *Proceedings of IEEE Workshop on Multimedia Signal Processing*, pp. 161–164, 2002.

[55] Han B., Ji H., and Gao X., A 3D wavelet and motion vector based method for wipe transition detection. In *Proceedings of ICSP*, Vol. 2, pp. 1207–1210, 2004.

[56] Mackowiak S. and Relewicz M., Wipe transition detection based on motion activity and dominant colors descriptors. In *Proceedings of 4th International Symposium on Image Signal Processing and Analysis*, pp. 480–483, 2005.

[57] Li Yufeng, Xang Yinghua, and Li Guiju, A novel wipe transition detection based on multi-feature. In *Proceedings of the Third International Conference on Knowledge Discovery and Data Mining*, 2010.

[58] Truong B.T. and Venkatesh S., Determining dramatic intensification via flashing lights in movies. In *Proceedings of IEEE International Conference on Multimedia and Expo*, pp. 60–63, 2001.

[59] Yuliang G. and De X., A solution to illumination variation problems in shot detection. In *Proceedings of TENCON 2004, IEEE Region 10 Conference*, Vol. 2, pp. 81–84, 2004.

[60] Qian X., Liu G., and Su R, Effective fades and flashlight detection based on accumulating histogram difference. *IEEE Transactions on Circuits and Systems for Video Technology*, 16(10):1245–1258, 2006.

[61] Fernando W.A.C., Canagarajah C.N., and Bull D.R., Fade and dissolve detection in uncompressed and compressed video sequences. In *Proceedings of IEEE, ICIP*, pp. 299–303, 1999.

[62] Truong B.T., Dorai C., and Venkatesh S., New enhancements to cut, fade, and dissolve detection processes in video segmentation. In *ACM Multimedia*, pp. 219–227, 2000.

[63] Lienhart R. and Zaccarin A., A system for reliable dissolve detection in video. In *Proceedings of IEEE ICIP*, Greece, pp. 406–409, 2001.

[64] Volkmer T., Tahaghoghi S.M.M., and Williams H.E., Gradual transition detection using average frame similarity. In *Proceedings of IEEE Computer Society Conference on Computer Vision and Pattern Recognition, CVPRW04*, 2004.

[65] Cai C., Lam K.M., and Tan Z., An efficient scene break detection based on linear prediction. In *Proceedings of International Symposium on Intelligent Multimedia, Video and Speech Processing*, Hongkong, pp. 555–558, 2004.

[66] Han B., Xinbo G., and Ji H., A unified framework for shot boundary detection. In *Proceedings of CIS2005*, Part I, Y. Hao et al. (Eds.), Lecture Notes in Artificial Intelligence, Vol. 3801, pp. 997–1002. Springer-Verlag, 2005.

[67] Ling J., Zhuang Y-T. and Lian Y-Q., A new method for shot gradual transition detetcion using support vector machine. In *Proceedings of the Fourth International Conference on Machine Learning and Cybernetics*, Guangzhou, pp. 5599–5604, 2005.

[68] Pedro J. S., Domínguez S., and Denis N., On the use of entrophy series for fade detection. In *Proceedings of CAEPIA2005*, R. Marín et al. (Eds.), Lecture Notes in Artificial Intelligence, Vol. 4177, pp. 360–369. Springer-Verlag, 2005.

[69] Bezerra F.N. and Leite N.J., Using string matching to detect video transitions. *Pattern Anal. Application*, 10:45–54, 2007.

[70] Liang B., Song L., Hai L., and Tiang B, Video shot boundary using Petri-net. In *Proceedings of the Seventh International Conference on Machine Learning and Cybernetics*, 2008.

[71] Li Jun, Ding Y., Y Shi Y., and Zeng Q., DWT based shot boundary detection using support vector machine. In *Proceedings of the Fifth International Conference on Information Assurance and Security*, 2009.

[72] Wenzhu Xu and Lihang Xu, A novel shot detection algorithm based on graph theory. In *Proceedings of the Second International Conference on Computer Engineering and Technology*, 2010.

[73] Padalkar M. and Zaveri M., Dissolve detection based on shot identification using singular value decomposition. In *Proceedings of the Fourth International Conference on Mathematical/Analytical Modelling and Computer Simulation*, 2010.

[74] Ewerth R. and Freisleben B., Unsupervised detection of gradual video shot changes with motion-based false alarm removel. In *Proceedings of ACIVS 2009*, Lecture Notes in Compuer Science, Vol. 5807. Springer-Verlag, 2009.

[75] Vasileios C., Aristidis L., and Nikolaos G., Simultaneous detection of abrupt cuts and dissolves in video using support vector machines. *Pattern Recognition Letters*, 30, 2009.

[76] Zhang H.J., Low C.Y., Smoliar S.W., and Tan S.Y., Video parsing and browsing using compressed data. *Multimedia Tools and Application*, 1:89–111, 1995.

[77] Onur K., Uğur G, and OzgUr U., Fuzzy color histogram-based video segmentation. *Computer Vision and Image Understanding*, 114, 2010.

[78] Ares M.E. and Barreiro A., Using a rank fusion technique to improve shot boundary detection effectiveness. *EUROCAST 2009*, Lecture Notes in Computer Science, Vol. 5717. Springer-Verlag, 2009.

[79] Arman F., Hsu A., and Lee M.Y., Image processing on compressed data for large video databases. In *Proceedings of International Conference on Multimedia*, Anaheim CA, pp. 267–272, 1993.

[80] Li Y-N., Lu Z-M., and Niu X-M., Fast video shot boundary detection framework employing pre-processing techniques. *IET Image Processing*, 3(3):121–134, 2009.

[81] Brojeshwar B. and Kaustav G., SVM based shot boundary detection using block motion feature based on statistical moments. In *Proceedings of the Seventh International Conference on Advances in Pattern Recognition*, 2009.

[82] Yu Meng, Li-Gong Wang, and Li-Zeng Mao, A shot boundary detection algorithm based on paticle swarm optimization classifier. In *Proceedings of the Eighth International Conference on Machine Learning and Cybernetics*, 2009.

[83] Li Xiuqiang, Xiao Guoqiang, Jiang Jianmin, Du Kuiran, and Qui Kaijin, Shot boundary detection based on SVMs via visual attention features. In *Proceedings of the International forum on Information Technology and Applications*, 2009.

[84] Barbu T., Novel automatic video cut detection technique using Gabor filtering. *Computers and Electrical Engineering*, 35, 2009.

[85] Sakarya U. and Telatar Z., Video scene detection using graph based representations. *Signal Processing:Image Communication*, vol. 25, 2010.

[86] Ruiloba R., Joly P., Marchand-Maillet S., and Quenot G., Towards a standard proptocol for the evaluation of video-to-shots segmentation algorithms. In *Proceedings of European Workshop on Content Based Multimedia Indexing*, Toulouse, 1999.

[87] Warhade K.K., Merchant S.N., and Desai U.B., Effective algorithm for detecting various wipe patterns and discriminating wipe from object and camera motion. *Image Processing, IET*, 4(6):429–442, 2010.

[88] Chung K., Chen T., and Yan W., New memory and computation efficient Hough transform for detecting lines. *Pattern Recognition*, 37:953–963, 2004.

[89] Fernandes A.F. and Oliverira M., Real time line detection through an improved Hough transform voting scheme. *Pattern Recognition*, 41:299–314, 2008.

[90] Ford R.M., Temple D., and Gerlach M., Metrics for shot boundary detection in digital video sequence. *Multimedia Systems*, 8:37–46, 2000.

[91] Warhade K.K., Merchant S.N., and Desai U.B., Shot boundary detection in the presence of flashlight. *Signal Image and Video Processing*, in press.

[92] Gonzalez R.C. and Woods R.E., *Digital Image Processing*. Addison-Wesley, 1992.

[93] Jorge L.C. and Sanz (Eds.), *Image Technology: Advances in Image Processing, Multimedia and Machine Vision*, pp. 361–383. Springer, Berlin, 1996.

[94] Warhade K.K., Merchant S.N., and Desai U.B., Avoiding false positive due to flashlights in shot detection using illumination suppression algorithm. In *Proceedings of the 5th IET International Conference on Visual Information Engineering*, Xian, China, pp. 377–382, 2008.

[95] Daubechies I., Ten lectures on wavelets. In *Society for Industrial and Applied Mathematics, CBMS-NSF*, Regional Conference Series in Applied Mathematics, Vol. 61, 1992.

[96] Leszczuk M. and Papir Z., Accuracy vs. speed tradeoff in detecting of shots in video content for abstracting digital video libraries. In *Proceedings of IDMS/PROMS 2002*, Lecture Notes in Computer Sciences, Vol. 2515, pp. 176–189. Springer-Verlag.

[97] Warhade K.K., Merchant S.N., and Desai U.B., Shot boundary detection in the presence of fire flicker and explosion using stationary wavelet transform. *Signal Image and Video Processing*, DOI 10.1007/s11760-010-0163-y, 2010.

[98] Mallat S.G., A theory for multiresolution signal decomposition: The wavelet representation. *IEEE Transaction on Pattern Analysis and Machine Intelligence*, 11(7):674–693, 1989.

[99] Nason G.P. and Silverman B.W., The stationary wavelet transform and some statistical applications. In *Wavelets and Statistics*, E. Antoniadis and G. Oppenheim (Eds.), pp. 281–299. Springer Verlag, 1995.

[100] Ogden R.T., *Essential Wavelets for Statistical Application and Data Analysis*, Birkhauser, New York, 1995.

[101] Warhade K.K., Merchant S.N., and Desai U.B., Performance evaluation of shot boundary detection metrics in the presence of object and camera motion. *IETE Journal of Research*, November 2011.

[102] Wang Z., Bovik A.C., Sheikh H.R., and Simoncelli E.P., Image quality assessment: From error visibility to structural similarity. *IEEE Transactions on Image Processing*, 13(4):600–612, 2004.

[103] Warhade K.K., Merchant S.N., and Desai U.B., Shot boundary detection in the presence of illumination and motion by using dual tree complex wavelet trasnform. *Signal, Image and Video Processing*, in press.

[104] Kingsbury N.G., The dual tree complex wavelet transform: A new technique for shift invariance and directional filters. In *Proceedings of 8th IEEE DSP Workshop*, Utah, 1998.

[105] Kingsbury N.G., Image processing with complex wavelet. *Phil. Trans. Royal Soceity London A*, 357:2543–2560, 1999.

[106] Selenick I.W., The design of approximate Hilbert transform pairs of wavelet bases. *IEEE Trans. Signal Processing*, 50(5):1144–1152, 2002.

[107] Selenick I.W., Baraniuk R.G., and Kingsbury N.G., The dual tree complex wavelet transform: A coherent framework for multiscale signal and image processing. *IEEE Signal Processing Magazine*, 22(6):123–151, 2005.

[108] Wang Z. and Simoncelli E.P., Translation insensitive image similarity in complex wavelet domain. In *Proceedings of IEEE International Conference on Acoustic, Speech and Signal Processing*, Vol. II, pp. 573–576, 2005.

Index

About the Authors

Krishna K. Warhade received his Bachelor of Engineering in Electronics in 1995 and Master of Engineering in Instrumentation in 1999 both from Shri Guru Gobind Singhji Institute of Engineering and Technology, Nanded, and Ph.D. in 2010 from the Department of Electrical Engineering, Indian Institute of Technology Bombay, India. He has 15 years of experience in teaching and research. He is currently working as a Professor in the Department of Electronics Engineering, Lokmanya Tilak College of Engineering, affiliated to University of Mumbai, India. His research interests are in the area of signal processing, image processing, video segmentation, video retrieval and wavelets. Currently he is a post-doctoral researcher in the SPANN Lab, Department of Electrical Engineering, IIT Bombay.

Shabbir N. Merchant is a Professor in Department of Electrical Engineering, IIT Bombay. He received his B. Tech., M. Tech. and Ph.D. degrees all from Department of Electrical Engineering, Indian Institute of Technology Bombay, India. He has more than 25 years of experience in teaching and research. Dr. Merchant has made significant contributions in the field of signal processing and its applications. His noteworthy contributions have been in solving state of the art signal and image processing problems faced by Indian defence. His broad area of research interests are signal and image processing, multimedia communication, wireless sensor networks and wireless communications, and has published extensively in these areas. He has been a chief investigator for a number of sponsored and consultancy projects. He has served as a consultant to both private industries and defence organizations. Dr. Merchant is a reviewer for many leading international and national journals and conferences. He is a Fellow of IETE. He is recipient of the 10th IETE Prof. S.V.C. Aiya Memorial Award for his contribution in the field of detection and tracking. He is also a recipient of 9th IETE SVC Aiya Memorial Award for 'Excellence in Telecom Education'.

Uday B. Desai received the B. Tech. degree from the Indian Institute of Technology, Kanpur, India in 1974, the M.S. degree from the State University of New York, Buffalo, in 1976, and the Ph.D. degree from the Johns Hopkins University, Baltimore, USA in 1979, all in Electrical Engineering. Since June 2009, Professor Desai is in charge as the first Director of IIT Hyderabad. From 1979 to 1984 he was as Assistant Professor in the Electrical Engineering Department at Washington State University, Pullman, WA, USA, and an Associate Professor at the same place from 1984 to 1987. From 1987 to 2009 he was a Professor in the Department of Electrical Engineering, Indian Institute of Technology Bombay, India. He was Dean of students at IIT-Bombay from August 2000 to July 2002. He has held Visiting Associate Professor's position at Arizona State University, Purdue University, and Stanford University. He was a Visiting Professor at EPFL, Lausanne during the summer of 2002. From July 2002 to June 2004 he was the Director of HP-IITM R and D Lab. at IIT-Madras. Professor Desai is Fellow of Indian National Science Academy (INSA), Fellow of Indian National Academy of Engineering (INAE), and recipient of the JC Bose Fellowship. His research interests are in wireless sensor networks, cognitive radio, signal and image processing and broadly in the area of cyber physical systems.